Paul M. Gauthier

Lectures on Several Complex Variables

 Birkhäuser

Paul M. Gauthier
Départment de Mathématiques et de Statistique
Université de Montréal
Montreal, QC, Canada

ISBN 978-3-319-37604-2   ISBN 978-3-319-11511-5 (eBook)
DOI 10.1007/978-3-319-11511-5
Springer Cham Heidelberg New York Dordrecht London

Mathematics Subject Classification (2010): 3201

Printed on acid-free paper

Springer is part of Springer Science+Business Media (www.springer.com)

Preface

This is a slightly amplified English translation of a course given at Université de Montréal in 2004. Prerequisites for the course are functions of one complex variable and functions of several real variables and topology, all at the undergraduate level. A previous encounter with subharmonic functions in the complex plane would be helpful, but the reader may consult introductory texts on complex analysis for such material. We deliberately require the student to fill in many details, but we indicate precisely when this is the case. Some important theorems are stated as problems. These missing details and proofs are carefully orchestrated to be feasible and instructive in context, and the course in this format was a success. These notes give only a short and swift overview of concepts and naturally indicate my own preferences, so instructors could use these notes as an indicator and then build their lectures of their own liking, with additional examples and exercises. While all of the textbooks in the bibliography are excellent, I wish to point out the unusual nature of the book by Kaplan [7]. Kaplan's book is one of the rare books that presents several complex variables at the undergraduate level and yet manages to present a significant amount of important material. Moreover, it has many elementary exercises and excellent illustrations. While the present notes are aimed primarily at graduate students, more ambitious undergraduate students and research mathematicians whose specialty is not several complex variables may also benefit from these lecture notes. It is my hope that this introduction to complex analysis in several variables, though brief, will be sufficient to inspire the student who has gone through them to have the curiosity to attend many interesting colloquia and seminars that touch upon topics related to those herein presented.

It is a pleasure to thank our friend Michael Range for helpful suggestions.

Montreal, QC, Canada   Paul M. Gauthier

Contents

Chapter 1
Introduction

Abstract In this introductory chapter, important and surprising differences between one and several complex variables are pointed out.

By way of introduction, we point out some major differences between one and several complex variables. The uniformization theorem, in a single complex variable, states that simply connected domains are biholomorphic to the disc, the plane, or the Riemann sphere, so every domain has one of these three as universal cover. In several variables, the natural generalization of the complex plane is complex Euclidean space $\mathbb{C}^n$ and the natural generalization of the Riemann sphere is complex projective space $\mathbb{P}^n$, but there is no single natural generalization of the unit disc $\mathbb{D}$. Two natural generalizations are the unit polydisc $\mathbb{D}^n$ and the unit ball $\mathbb{B}^n$. But these two domains are not biholomorphically equivalent [12]. In fact, in several variables, there are many nonequivalent simply connected domains. Many properties of the unit disc are shared by convex domains, but linear convexity (even in a single variable) is not biholomorphically invariant. In several variables, there appears a variant type of convexity which is biholomorphically invariant. This different type is the so-called pseudoconvexity, and it behaves like a linear convexity with respect to finding local minima. The class of open subsets of a complex Euclidean space, characterized by this property of pseudoconvexity, has become the most important subject of investigation in several complex variables. Such an open set is characterized by the property of being the natural domain of definition of some holomorphic function.

In a single complex variable, every domain is the natural domain of some holomorphic function. That is, for every domain $\Omega \subset \mathbb{C}$, there is a holomorphic function in Ω, which is holomorphic in no larger domain (or even larger Riemann surface). In several variables, the situation is dramatically different. Let $\Omega \subset \mathbb{C}^n, n \geq 2$, be a bounded domain with connected boundary $\partial\Omega$. Then, every function holomorphic on a neighborhood of $\partial\Omega$ extends to a function holomorphic on Ω. This result, introduced by Hartogs in 1906 and known as the Hartogs phenomenon, is considered by some to be the beginning of the theory of several complex variables. Hartogs' original argument was quite vague. It was believable but not a proof. A complete and correct proof was given in 1940/1941 by Martinelli

© Springer International Publishing Switzerland 2014
P.M. Gauthier, *Lectures on Several Complex Variables*,
DOI 10.1007/978-3-319-11511-5_1

and Bochner via the Bochner–Martinelli integral formula (see, for example, [13]). In 2007 Merkel and Porten [10] precisely filled in the gaps in Hartogs' original argument.

Another striking aspect of functions of several complex variables is also due to Hartogs. It is the Hartogs theorem on separate analyticity. A simplified version states that, if $f : \mathbb{C}^2 \to \mathbb{C}$ has the property that, for each $z \in \mathbb{C}$, $f(z, w)$ is an analytic function of w and, for each $w \in \mathbb{C}$, $f(z, w)$ is an analytic function of z, then f is analytic. In concise form, for functions of several complex variables, separate analyticity implies joint analyticity. Such is not the case for real-analyticity as the following problem illustrates.

Problem 1. The function

$$f(x, y) = \begin{cases} \frac{xy}{x^2+y^2} & xy \neq 0, \\ 0 & xy = 0. \end{cases}$$

is real-analytic on each line in the (x, y) plane, but is not even continuous.

A much stronger example was given [2] of a discontinuous function in $\mathbb{R}^2$, which is real-analytic even on each *analytic arc* in the (x, y) plane!

The student should be aware that, although we have pointed out some important differences between complex analysis in a single variable and complex analysis in n variables, for $n > 1$, many properties and proofs are the same for one or for several variables.

Chapter 2
Basics

Abstract Two striking theorems of Hartogs, which have no counterpart in a single variable, are introduced: the Hartogs theorem on separate holomorphy and the Hartogs phenomenon on the possibility of extending holomorphic functions.

The coordinates of $z \in \mathbb{C}^n$ are given by $z = (z_1, \cdots, z_n)$, and we write $z_j = x_j + i y_j$, where $x_j = \Re z_j$ and $y_j = \Im z_j$,

$$x = (x_1, \cdots, x_n) = \Re z, \quad y = (y_1, \cdots, y_n) = \Im z.$$

We denote the standard Hermitian inner product

$$\langle z, w \rangle = \sum_{j=1}^{n} z_j \overline{w}_j,$$

the norm

$$|z| = \langle z, z \rangle^{1/2} = \sqrt{\sum |z_j|^2} = \sqrt{\sum (|x_j|^2 + |y_j|^2)} = \sqrt{|x|^2 + |y|^2}$$

and the differential operators

$$\frac{\partial}{\partial z_j} = \frac{1}{2} \left(\frac{\partial}{\partial x_j} - i \frac{\partial}{\partial y_j} \right), \quad \frac{\partial}{\partial \overline{z}_j} = \frac{1}{2} \left(\frac{\partial}{\partial x_j} + i \frac{\partial}{\partial y_j} \right).$$

When speaking, we call these the derivative with respect to z_j and the derivative with respect to $\overline{z}_j$, respectively; however, they are not necessarily derivatives. That is, for a C^1 function f, the expressions $\frac{\partial f}{\partial z_j}$ and $\frac{\partial f}{\partial \overline{z}_j}$ are well defined above, but cannot always be expressed as the limit of some differential quotient.

Recall that a complex-valued function f defined in an open subset Ω of $\mathbb{C}$ is said to be holomorphic if f has a derivative at each point of Ω. If f is holomorphic in an open set Ω of $\mathbb{C}$, then f satisfies the Cauchy–Riemann equation $\partial f / \partial \overline{z} = 0$ in Ω. The converse is false. For example, the function f defined to be 0 at 0 and

P.M. Gauthier, *Lectures on Several Complex Variables*,
DOI 10.1007/978-3-319-11511-5_2

e^{-1/z^4} elsewhere satisfies the Cauchy–Riemann equation at all points of $\mathbb{C}$ but is not holomorphic at 0. However, if $f \in C^1(\Omega)$, then f is holomorphic in Ω if and only if it satisfies the Cauchy–Riemann equation.

Let Ω be an open subset of $\mathbb{C}^n$. A function $f \in C^1(\Omega)$ is said to be *holomorphic* in Ω if it is holomorphic in each variable, thus, if and only if f satisfies the system of (homogeneous) Cauchy–Riemann equations

$$\frac{\partial f}{\partial \bar{z}_j} = 0, \ j = 1, \cdots, n.$$

It is easily verified that the family $\mathcal{O}(\Omega)$ of holomorphic functions on an open set $\Omega \subset \mathbb{C}^n$ is an algebra over $\mathbb{C}$.

As usual, by a *domain* in $\mathbb{C}^n$, we mean an open connected set. Since the topology of $\mathbb{C}^n$ is the same as that of the underlying real Euclidean space $\mathbb{R}^{2n}$, an open subset of $\mathbb{C}^n$ is connected if and only if it is pathwise connected. It is also easy to see that if $\Omega \subset \mathbb{C}^n$ is a domain, then every two points in Ω can be connected by a polygonal path each of whose segments lies in a complex line in a coordinate direction. That is, along each segment, only one of the n variables $z_1, \cdots, z_n$ varies. To see this, fix a point $p \in \Omega$ and let U be the set of points in Ω which can be attained from p by such a path. It is easy to see that U is both open and closed in Ω and since Ω is connected, $U = \Omega$.

Theorem 1 (Uniqueness). *If f is holomorphic in a domain $\Omega \subset \mathbb{C}^n$, and $f = 0$ in a (nonempty) open subset $U \subset \Omega$, then $f = 0$ on Ω.*

Proof. Let V be the set of points p in Ω such that $f = 0$ in a neighborhood of p. Using the fact that the theorem is true for $n = 1$ and that every two points in Ω can be joined by a polygonal path each segment lies in a complex line in a coordinate direction, it is easy to see that V is open and closed in Ω. Since Ω is connected, it follows that $V = \Omega$.   □

Corollary 2 (Uniqueness). *Let f and g be holomorphic in a domain Ω and suppose $f = g$ on an open subset of Ω. Then $f = g$ on Ω.*

Corollary 3 (Maximum modulus principle). *If f is holomorphic in a domain Ω and $|f|$ attains a local maximum at a point $p \in \Omega$, then f is constant.*

Proof. Let f be holomorphic in a domain Ω and suppose $|f|$ attains a local maximum at a point $p \in \Omega$. If in a neighborhood of p we fix $n - 1$ coordinates and let the other coordinate, call it z_j, vary, we obtain a holomorphic function $g(z_j)$, which, by the maximum principle for functions of one variable, is constant for z_j near p_j. Since we can do this with each variable, we obtain that f is constant near p along each complex line passing through p in a coordinate direction. If ℓ is such a line, one-variable arguments show that f is constant on the component of $\ell \cap \Omega$ containing p. Let V be a connected open neighborhood of p in which $|f(p)|$ is maximal. We can join p to any point $q \in V$ by a finite chain of discs lying in the intersection $\ell \cap V$ of complex lines ℓ in the coordinate directions. Applying the above argument along this chain of discs, we conclude that $f(p) = f(q)$. Thus, f is constant in the neighborhood V of p. Consider the function $g = f - f(p)$. Since $g \in \mathcal{O}(\Omega)$ and $g = 0$ on the open subset V, it follows from the uniqueness principle that $g = 0$ everywhere in Ω. That is, f is constant. $\square$

If f is holomorphic in $\mathbb{C}^n$, we say that f is an *entire* function of n complex variables. As an example, if φ is an entire function of one complex variable, then $f(z, w) \equiv \varphi(z)$ is easily seen to be an entire function of two complex variables.

It is a deep result of Hartogs that the condition that f be in $C^1(\Omega)$ is superfluous. Thus, if Ω is an open subset of $\mathbb{C}^n$ and $f : \Omega \to \mathbb{C}$ has the property that, for each fixed $c_1, \cdots, c_{j-1}, c_{j+1}, \cdots, c_n$, the function

$$f(c_1, \cdots, c_{j-1}, z_j, c_{j+1}, \cdots, c_n)$$

is a holomorphic function of the single variable z_j, in

$$\{z_j \in \mathbb{C} : (c_1, \cdots, c_{j-1}, z_j, c_{j+1}, \cdots, c_n) \in \Omega)\},$$

then $f \in \mathcal{O}(\Omega)$. Note that there is no regularity assumption on f; it is not even assumed to be measurable!

A function is said to be *holomorphic* on a subset E of $\mathbb{C}^n$ if it is holomorphic in an open neighborhood of E. In complex analysis, the inhomogeneous system of Cauchy–Riemann equations

$$\frac{\partial f}{\partial \bar{z}_j} = u_j, \ j = 1, \cdots, n,$$

is also important. Loosely speaking, we say that a system of differential equations is *integrable* if the system has a solution. Of course, in order for a solution to exist to the above inhomogeneous system, the functions u_j must satisfy the following integrability (or compatibility) conditions:

$$\frac{\partial u_j}{\partial \bar{z}_k} = \frac{\partial u_k}{\partial \bar{z}_j} \quad j, k = 1, \cdots, n.$$

A function defined in an open subset of $\mathbb{R}^n$ (respectively $\mathbb{C}^n$) is said to be real (respectively complex) *analytic* if it is locally representable by power series.

Theorem 4. *A function is holomorphic if and only if it is complex analytic.*

Theorem 5. *A function is complex analytic if and only if it is complex analytic in each variable.*

Problem 2. Show Theorem 4 implies Theorem 5.

Problem 3. Show that the "real" analog of Theorem 5 is false. This is a big difference between real analysis and complex analysis.

Let f_j be holomorphic in an open connected subset Ω_j of $\mathbb{C}^n$, $j = 1, 2$ with $\Omega_1 \cap \Omega_2 \neq \emptyset$ and suppose $f_1 = f_2$ in some nonempty component G of $\Omega_1 \cap \Omega_2$. Then f_2 is said to be a *direct holomorphic continuation* of f_1 through G. In shorthand, we also say (f_2, Ω_2) is a direct holomorphic continuation of (f_1, Ω_1).

Let f be holomorphic in a domain Ω and let $p \in \partial\Omega$. We say that f has a *direct holomorphic continuation to p* if there is a holomorphic function f_p in a neighborhood U_p of p such that (f_p, U_p) is a direct holomorphic continuation of (f, Ω) through some component G of $\Omega \cap U_p$ with $p \in \partial G$.

Problem 4. In $\mathbb{C}$ give an example of a function f holomorphic in a domain D and a boundary point p such that f has a direct holomorphic continuation to p. Also, give an example where f has no direct holomorphic continuation to p.

A domain Ω is a *domain of holomorphy* if it is the "natural" domain for some holomorphic function, that is, if there is a function f holomorphic in Ω which cannot be directly holomorphically continued to any boundary point of Ω. In particular, f cannot be directly holomorphically continued to any domain which contains Ω.

Problem 5. Give an example of a domain of holomorphy in $\mathbb{C}^1$.

Problem 6. Show that each domain in $\mathbb{C}^1$ is a domain of holomorphy.

Problem 7. Give an example of a domain of holomorphy in $\mathbb{C}^2$.

The following theorem and its corollary show an important difference between complex analysis in one variable and in several variables.

Theorem 6 (Hartogs phenomenon). *Let Ω be a bounded domain in $\mathbb{C}^n$, $n > 1$, with connected boundary. Then, any function holomorphic in a neighborhood of $\partial\Omega$ has a direct holomorphic continuation to Ω.*

Corollary 7. *In $\mathbb{C}^n$, $n > 1$, not every domain is a domain of holomorphy.*

Problem 8. Show that the corollary follows from the theorem.

Corollary 8. *Holomorphic functions of more than one variable have no isolated nonremovable singularities.*

Problem 9. Show that the corollary follows from the theorem.

Corollary 9. *Holomorphic functions of more than one variable have no isolated zeros.*

Problem 10. Show that the corollary follows from the previous corollary.

In $\mathbb{C}$ there are two domains of particular interest, $\mathbb{C}$ and the unit disc $\mathbb{D}$. The Riemann mapping theorem asserts that each simply connected domain in $\mathbb{C}$ is equivalent, in the sense of complex analysis, to one of these two domains.

In $\mathbb{C}^n$, the analog of the Riemann mapping theorem fails. First of all, there are at least two obvious and natural generalizations of the unit disc, the unit ball $\mathbb{B}^n = \{z : |z| < 1\}$ and the unit polydisc $\mathbb{D}^n = \{z : |z_j| < 1, j = 1, \cdots, n\}$. Both of these domains are simply connected, but they are *not* equivalent in the sense of complex analysis. Let us be more precise.

A mapping from a domain of $\mathbb{C}^n$ into $\mathbb{C}^m$ is said to be *holomorphic* if each of its components is holomorphic. A holomorphic mapping from one domain to another is said to be biholomorphic if it is bijective and if the inverse mapping is also holomorphic. The two domains are then said to be biholomorphically equivalent. Poincaré has shown (see [9]) that, for $n > 1$, the unit polydisc $\mathbb{D}^n$ and the unit ball $\mathbb{B}^n$ are not biholomorphic!

The Hartogs phenomenon and the failure of the Riemann mapping theorem, for $n > 1$, are two major differences between complex analysis in one variable and in several variables.

Chapter 3
Cauchy Integral Formula

Abstract Holomorphic functions on a polydisc are represented by the Cauchy integral of their values on the distinguished boundary of the polydisc.

The following is the Cauchy formula for the polydisc.

Theorem 10. *Let f be holomorphic on the closed unit polydisc $\overline{\mathbb{D}}^n$. Then,*

$$\frac{1}{(2\pi i)^n} \int_{|\zeta_1|=1} \int_{|\zeta_2|=1} \cdots \int_{|\zeta_n|=1} \frac{f(\zeta_1, \zeta_2, \ldots, \zeta_n)}{\prod_{j=1}^{n}(\zeta_j - z_j)} d\zeta_1 d\zeta_2 \ldots d\zeta_n$$

for each $z = (z_1, z_2, \ldots, z_n) \in \mathbb{D}^n$.

Proof. For simplicity of notation, we shall give the proof only for $n = 2$. For each fixed z_2 in the unit disc, $f(z_1, z_2)$ is holomorphic in z_1 for z_1 in the closed unit disc. Hence, for $|z_1| < 1$,

$$f(z_1, z_2) = \frac{1}{2\pi i} \int_{|\zeta_1|=1} \frac{f(\zeta_1, z_2)}{\zeta_1 - z_1} d\zeta_1$$

by the usual Cauchy formula. For each fixed ζ_1 on the unit circle, $f(\zeta_1, z_2)$ is holomorphic in z_2 for z_2 in the closed unit disc. Hence, for $|z_2| < 1$,

$$f(\zeta_1, z_2) = \frac{1}{2\pi i} \int_{|\zeta_2|=1} \frac{f(\zeta_1, \zeta_2)}{\zeta_2 - z_2} d\zeta_2.$$

Combining the last two expressions, we obtain the theorem. □

The above theorem is the Cauchy Formula for polydiscs. We have stated it for the unit polydisc, but it holds for polydiscs in general. It says that a function f holomorphic in a neighborhood of a closed polydisc $\overline{\mathcal{D}}$ can be expressed in the polydisc as its Cauchy integral over a portion of the boundary $\partial\mathcal{D}$ which is just the cartesian product of circles. This portion of the boundary of the polydisc is called the *distinguished boundary*. In particular, the values of f in $\mathcal{D}$ are completely determined by its values on the distinguished boundary. Let us compare this with the one-variable situation. A holomorphic function on the closed unit disc $\overline{\mathbb{D}}^1$ is

© Springer International Publishing Switzerland 2014
P.M. Gauthier, *Lectures on Several Complex Variables*,
DOI 10.1007/978-3-319-11511-5_3

completely determined inside $\mathbb{D}^1$ by values on the entire boundary ∂D^1. In this sense, the distinguished boundary is the important part of the boundary. However, for $n > 1$, the distinguished boundary is a small portion of the boundary. *Its real dimension is n while the real dimension of the entire boundary of the polydisc is $2n - 1$.*

Problem 11. Show that each function holomorphic in the polydisc is the uniform limit on compact subsets of rational functions.

Problem 12. Show that each entire function (function holomorphic in $\mathbb{C}^n$) is the uniform limit on compact subsets of rational functions.

Problem 13. Extend Theorem 10 to the case where f is holomorphic on a closed polydisc with not necessarily equal radii.

Later, we shall show that functions holomorphic in a polydisc have a power series representation, from which it follows that they can in fact be approximated by polynomials.

We have stated the Cauchy formula in the polydisc in $\mathbb{C}^n$. As in one variable, there is also a Cauchy integral formula for derivatives. To state the formula in $\mathbb{C}^n$, we introduce multi-index notation $\alpha = (\alpha_1, \cdots, \alpha_n)$, where each α_j is a nonnegative integer and, by abuse of notation, we write $1 = (1, \cdots, 1)$ and $0 = (0, \cdots, 0)$. If $a \in \mathbb{C}^n$ and $a_j \neq 0, j = 1, \cdots, n$, we write

$$\frac{z}{a} = \frac{z^1}{a^1} = \frac{z_1 \cdots z_n}{a_1 \cdots a_n}. \tag{3.1}$$

Set $|\alpha| = \alpha_1 + \cdots + \alpha_n$, $\alpha! = \alpha_1! \cdots \alpha_n!$ and $z^\alpha = z_1^{\alpha_1} \cdots z_n^{\alpha_n}$. We denote derivatives with respect to real variables by

$$\frac{\partial^{|\beta|+|\gamma|} f}{\partial x^\beta \partial y^\gamma} = \frac{\partial^{|\beta|+|\gamma|} f}{\partial x_1^{\beta_1} \cdots \partial x_n^{\beta_n} \partial y_1^{\gamma_1} \cdots \partial y_n^{\gamma_n}}$$

and with respect to complex variables by

$$f^{(\alpha)} = \frac{\partial^{|\alpha|} f}{\partial z^\alpha} = \frac{\partial^{|\alpha|} f}{\partial z_1^{\alpha_1} \cdots \partial z_n^{\alpha_n}}.$$

The following lemma of Leibniz (see [1]) allows one to differentiate under the integral sign.

Lemma 11 (Leibniz). *Let μ be a measure on a locally compact Hausdorff space Y with countable base and let I be an open interval. Consider a function $f : I \times Y \to \mathbb{R}$, with $f(x, \cdot)$ (Borel) measurable, for each $x \in I$. Suppose there exists a point x_0 such that $f(x_0, \cdot)$ is μ-integrable, $\partial f/\partial x$ exists on I, and there is a μ-integrable function g on Y such that*

$$\left|\frac{\partial f}{\partial x}(x, y)\right| \le g(y), \quad \forall (x, y).$$

Then

$$\frac{\partial}{\partial x} \int f(x, \cdot) d\mu = \int \frac{\partial f}{\partial x}(x, \cdot) d\mu.$$

Let $b\mathbb{D}^n = \{z : |z_j| = 1, j = 1, \cdots, n\}$ denote the *distinguished boundary* of the unit polydisc and $d\zeta = d\zeta_1 \cdots d\zeta_n$.

Theorem 12. *Let f be holomorphic on the closed polydisc $\overline{\mathbb{D}}^n$. Then, $f \in C^\infty(\mathbb{D}^n)$, and for each $z \in \mathbb{D}^n$*

$$\frac{\partial^{|\beta|+|\gamma|} f}{\partial x^\beta \partial y^\gamma}(z) = \frac{1}{(2\pi i)^n} \int_{b\mathbb{D}^n} f(\zeta) \frac{\partial^{|\beta|+|\gamma|}}{\partial x^\beta \partial y^\gamma}\left(\frac{1}{\zeta - z}\right) d\zeta.$$

All of these partial derivatives are holomorphic and, in particular,

$$f^{(\alpha)}(z) = \frac{\alpha!}{(2\pi i)^n} \int_{b\mathbb{D}^n} \frac{f(\zeta)}{(\zeta - z)^{\alpha+1}} d\zeta.$$

Proof. We already have the Cauchy integral formula for f itself, that is, for the multi-index $\alpha = 0$. In order to obtain the Cauchy formula for the first-order partial derivatives of f, we apply the Leibniz theorem to differentiate the Cauchy formula for f by differentiating under the integral sign. Repetition of this process gives the general formula. We note that from this general Cauchy integral formula, it follows that all of the partial derivatives are continuous. Since all partial derivatives of the Cauchy kernel are holomorphic, the Leibniz formula yields that all partial derivatives of f are also holomorphic. The second formula is then a particular case of the first, since

$$\frac{\partial}{\partial z_j} = \frac{1}{2}\left(\frac{\partial}{\partial x_j} - i \frac{\partial}{\partial y_j}\right).$$

□

Problem 14. If f is holomorphic in an open set Ω of $\mathbb{C}^n$, then $f \in C^\infty(\Omega)$.

Problem 15. If f is holomorphic in an open set Ω of $\mathbb{C}^n$, then all partial derivatives of f are also holomorphic in Ω.

Theorem 13. *If f is a holomorphic function on a polydisc*

$$D(a, r) = \{z : |z_j - a_j| < r_j, j = 1, \cdots, n\},$$

which for simplicity we denote by D, *then*

$$|f(a)| \leq \frac{1}{\pi^n (r_1 \cdots r_n)^2} |f|_1$$

where

$$|f|_1 = \int_D |f| dV$$

is the integral with respect to Lebesgue measure in $\mathbb{C}^n$.

Proof. For simplicity of notation, we give the proof for $n = 2$. By the Cauchy formula, for each $0 < \rho < r$,

$$f(a) = \frac{1}{(2\pi i)^2} \int_{|\zeta_1 - a_1| = \rho_1} \int_{|\zeta_2 - a_2| = \rho_2} \frac{f(\zeta_1, \zeta_2)}{(\zeta_1 - a_1)(\zeta_2 - a_2)} d\zeta_1 d\zeta_2.$$

Thus

$$|f(a)| \leq \frac{1}{(2\pi)^2} \int_0^{2\pi} \int_0^{2\pi} |f(a_1 + \rho_1 e^{i\theta_1}, a_2 + \rho_2 e^{i\theta_2})| d\theta_1 d\theta_2.$$

Multiplying by ρ_1 and integrating with respect to ρ_1 from 0 to r_1 and writing dA for area measure,

$$\frac{r_1^2}{2} |f(a)| \leq \frac{1}{(2\pi)^2} \int_0^{2\pi} \int_0^{2\pi} \int_0^{r_1} |f(a_1 + \rho_1 e^{i\theta_1}, a_2 + \rho_2 e^{i\theta_2})| \rho_1 d\rho_1 d\theta_1 d\theta_2 =$$

$$\frac{1}{(2\pi)^2} \int_0^{2\pi} \int_{|\zeta_1 - a_1| < r_1} |f(\zeta_1, a_2 + \rho_2 e^{i\theta_2})| dA d\theta_2.$$

Now, multiplying by ρ_2 and integrating with respect to ρ_2 from 0 to r_2,

$$\frac{r_1^2 r_2^2}{2^2} |f(a)| \leq \frac{1}{(2\pi)^2} \int_{|\zeta_2 - a_2| < r_2} \left(\int_{|\zeta_1 - a_1| < r_1} |f(\zeta_1, \zeta_2)| dA \right) dA =$$

$$\frac{1}{(2\pi)^2} \int_\Omega |f| dV = \frac{1}{(2\pi)^2} |f|_1.$$

$\square$

Theorem 14. *Let φ be continuous on the distinguished boundary of a polydisc $\mathcal{D}$ and define F as the Cauchy integral of φ:*

$$F(z) = \frac{1}{(2\pi i)^n} \int_{b\mathcal{D}} \frac{\varphi(\zeta)}{\zeta - z} d\zeta,$$

for $z \in \mathcal{D}$. Then, F is holomorphic in $\mathcal{D}$.

Proof. The proof is the same as that of the Cauchy integral formula. Using the Leibniz formula, we show that F is smooth and satisfies the Cauchy–Riemann equations.   □

Problem 16. Extend Th. 14. The same formula (for $n = 2$) defines a holomorphic function in each of the following four domains: $\Omega_1 = \{(z, w) : |z| < 1, |w| < 1\}$, $\Omega_2 = \{(z, w) : |z| > 1, |w| < 1\}$, $\Omega_3 = \{(z, w) : |z| < 1, |w| > 1\}$, and $\Omega_4 = \{(z, w) : |z| > 1, |w| > 1\}$. In the particular case $\varphi \equiv 1$ evaluate the function F in each one of $\Omega_i, i = 1, 2, 3, 4$.

An important difference between complex analysis in one variable and in several variables is the existence of domains which are not domains of holomorphy in $\mathbb{C}^n$, $n > 1$.

The following fundamental example was discovered by Hartogs. In studying this example, the student should draw the absolute value diagram associated to this figure. This can be found in any book on several complex variables, in particular in [7]. This diagram should also be drawn in class.

Theorem 15. *In $\mathbb{C}^2$, consider the domain*

$$H = \{z : |z_1| < 1/2, |z_2| < 1\} \cup \{z : |z_1| < 1, 1/2 < |z_2| < 1\}.$$

Every function holomorphic in the domain H extends to the unit polydisc $\mathbb{D}^2 = \{z : |z_j| < 1, j = 1, 2\}$.

Proof. Let f be holomorphic in H. Fix $1/2 < \delta < 1$. Then,

$$F(z_1, z_2) = \frac{1}{2\pi i} \int_{|\zeta| = \delta} \frac{f(z_1, \zeta)}{\zeta - z_2} d\zeta \tag{3.2}$$

defines a holomorphic function in the polydisc

$$\mathbb{D}_\delta = \{z : |z_1| < 1, |z_2| < \delta\}.$$

The proof that F is holomorphic is the same as that of Theorem 14, using the Leibniz formula, noting that $F(z) = \int K(z, \zeta) d\zeta$, where the kernel K is continuous and holomorphic in z in the domain $|z_1| < 1, |z_2| \neq \delta$. Since for $|z_1| < 1/2$ the function $f(z_1, \cdot)$ is holomorphic on $|z_2| < 1$, formula (3.2) implies that $F(z_1, z_2) = f(z_1, z_2)$ in the polydisc $|z_1| < 1/2, |z_2| < \delta$. The uniqueness property of holomorphic functions implies that $F = f$ on the intersection of H and this polydisc. Thus, F is a direct holomorphic continuation of f from H to the polydisc $\mathbb{D}_\delta$. Since the union of H and this polydisc is the unit polydisc $\mathbb{D}^2$, this concludes the proof.   □

This Hartogs figure H can be used as an example to reveal other differences between $\mathbb{C}$ and $\mathbb{C}^n$. If $\{z_j\}$ is a sequence of distinct points in $\mathbb{C}$ having no accumulation point and $\{w_j\}$ is an arbitrary sequence of complex numbers, then there is an interpolating entire function f on $\mathbb{C}$ for this data, that is, an entire function f such that $f(z_j) = w_j, j = 1, 2, \cdots$. In fact, one can do the same for an arbitrary domain $\Omega \subset \mathbb{C}$: given a sequence of distinct points in Ω having no accumulation point in Ω, there exists a holomorphic function in Ω taking preassigned values on this sequence.

In $\mathbb{C}^n, n > 1$, just as there are domains of holomorphy and domains which are not domains of holomorphy, there are domains where interpolation in the sense of the previous paragraph is possible and domains where it is not. In the Hartogs figure H, such interpolation is not always possible. Indeed, consider in $\mathbb{C}^2$ the sequence $z^j = (1/2 - 1/j, 1/2 + 1/j), j = 1, 2, \cdots$, which converges to the point $z^\infty = (1/2, 1/2)$. The sequence $\{z^j\}$ lies in the Hartogs figure H and the limit z^∞ lies in the polydisc $\mathbb{D}^2$. Set $w_j = (-1^j), j = 1, 2, \cdots$. Suppose there were a function f holomorphic in H such that $f(z^j) = (-1)^j, j = 1, 2, \cdots$. By Theorem 15, f extends holomorphically to the polydisc $\mathbb{D}^2$, which gives a contradiction, since f cannot be extended continuously to the point $z^\infty \in \mathbb{D}^2$. Thus, interpolation on discrete sequences is not always possible in the Hartogs figure H.

On the other hand, it is a fact that such interpolation is always possible in all of $\mathbb{C}^n$. We shall not show this. On the contrary, we shall comment on why this is not obvious. To accomplish such an interpolation, one might be tempted, given a discrete sequence $\{z_j\}$, to find a complex direction ℓ such that the projections α_j of the z_j on the line ℓ form a discrete sequence of distinct points in ℓ. Given an arbitrary sequence $\{w_j\}$ of complex numbers, on $\ell = \mathbb{C}$, one would have an entire function of one complex variable f_1, such that $f_1(\alpha_j) = w_j, j = 1, 2, \cdots$. Now, choose a nonzero $e_1 \in \ell$ and form a basis $(e_1, \cdots, e_n)$ of $\mathbb{C}^n$. Let $(\zeta_1, \cdots, \zeta_n)$ be the coordinates of a point z in this basis. We can define an entire function f in $\mathbb{C}^n$, by setting $f(z) = f_1(\zeta_1)$. The entire function f would be the desired interpolating entire function. However, the first step of this attempt fails. We suggest, as an exercise, that the student show, for $n > 1$, the existence of a discrete sequence of distinct points in $\mathbb{C}^n$, for which there is no complex line through the origin on which the projection of the sequence is discrete.

Chapter 4
Sequences of Holomorphic Functions

Abstract Properties of sequences of holomorphic functions of several variables are quite similar to those of holomorphic functions of a single variable.

The theory of sequences of holomorphic functions in several variables is similar to that in one variable and so this section could be seen as a review of the one-variable theory while noting that the arguments work in greater generality.

Theorem 16. *On an open set, the uniform limit of holomorphic functions is holomorphic.*

Proof. Let f_n be holomorphic on an open set Ω and suppose $f_n \to f$ uniformly. It is sufficient to show that f is holomorphic in each polydisc whose closure is contained in Ω. Let $\mathcal{D}$ be such a polydisc. From the uniform convergence, we have, for $z \in \mathcal{D}$,

$$f(z) = \lim f_j(z) = \lim \frac{1}{(2\pi i)^n} \int_{b\mathcal{D}} \frac{f_j(\zeta)}{\zeta - z} d\zeta = \frac{1}{(2\pi i)^n} \int_{b\mathcal{D}} \frac{f(\zeta)}{\zeta - z} d\zeta.$$

Thus, f is a Cauchy integral in $\mathcal{D}$ and hence f is holomorphic in $\mathcal{D}$. □

One of the most fundamental facts concerning numerical sequences is the Bolzano–Weierstrass theorem. Recall that a sequence of numbers $\{z_j\}$ is *bounded* if there is a number $M > 0$ such that $|z_j| \leq M$, for all j.

Theorem 17 (Bolzano–Weierstrass). *Any bounded sequence of numbers has a convergent subsequence.*

A sequence of functions $\{f_j\}$ is (uniformly) *bounded* on a set E if there is a number $M > 0$ such that $|f_j| \leq M$, for all j. For sequences of functions, we have the following analog of the Bolzano–Weierstrass theorem, known as Montel's theorem.

Theorem 18 (Montel). *Let $\mathcal{F}$ be a bounded family of holomorphic functions on an open set $\Omega \subset \mathbb{C}^n$. Then, each sequence of functions in $\mathcal{F}$ has a subsequence which converges uniformly on compact subsets.*

© Springer International Publishing Switzerland 2014
P.M. Gauthier, *Lectures on Several Complex Variables*,
DOI 10.1007/978-3-319-11511-5_4

In order to prove Montel's theorem, we gather a certain amount of material which is, in any case, interesting in itself.

Recall that a family $\mathcal{F}$ of complex-valued functions, defined on a metric space (X, d) is *equicontinuous* if for each $\epsilon > 0$, there is a $\delta > 0$ such that, for all $f \in \mathcal{F}$ and for all $p, q \in X$,

$$d(p, q) < \delta \quad \text{implies} \quad |f(p) - f(q)| < \epsilon.$$

Theorem 19 (Arzelà-Ascoli). *If K be a compact metric space and $\{f_j\}$ is a sequence of complex-valued functions which is pointwise bounded and equicontinuous on K, then*

(a) $\{f_j\}$ is uniformly bounded on K;
(b) $\{f_j\}$ has a uniformly convergent subsequence.

Problem 17. Use Theorem 12 to show that if $\mathcal{F}$ is a bounded family of holomorphic functions on an open subset $\Omega \subset \mathbb{C}^n$, then the family $\nabla \mathcal{F} = \{\nabla f : f \in \mathcal{F}\}$ is bounded on compact subsets of Ω.

Problem 18. Let f be a smooth function defined in an open convex subset B of $\mathbb{R}^n$. If $|\nabla f| \leq M$ in B, then $|f(p) - f(q)| \leq M|p - q|$, for each $p, q \in B$.

Let X be a topological space. An exhaustion of X by compact sets is a sequence $\{K_j\}$ of nested compact subsets, $K_j \subset K_{j+1}^0$, whose union is X.

Problem 19. Show that each open subset of $\mathbb{R}^n$ admits an exhaustion by compact sets.

Proof (of Montel theorem). Let $\{f_j\}$ be a bounded sequence of holomorphic functions on an open set $\Omega \subset \mathbb{C}^n$. Now, let K be a compact subset of Ω. Let d be the distance of K from $\partial \Omega$ and choose $2r < d$. We may cover K by finitely many balls $B(a_1, r) \cdots, B(a_m, r)$ whose centers are in K. From Problem 17, it follows that the sequence $\{\nabla f_j\}$ of gradients of the sequence $\{f_j\}$ is uniformly bounded on the union of the closed balls $\overline{B}(a_1, 2r), \cdots, \overline{B}(a_m, 2r)$ by some $M < +\infty$. If $z, \zeta \in K$ and $|z - \zeta| < r$, then, since z lies in some $B(a_k, r)$, both z and ζ lie in $B(a_k, 2r)$. By Problem 18, it follows that $|f_j(z) - f_j(\zeta)| \leq M|z - \zeta|, j = 1, 2, \cdots$. Thus, the sequence $\{f_j\}$ is equicontinuous on K. Since the sequence is also by hypothesis bounded on K, it follows from the Arzelá-Ascoli theorem that the sequence $\{f_j\}$ has a subsequence, which converges uniformly on K.

By Problem 19, the open set Ω has an exhaustion by compact sets:

$$K_1 \subset K_2^0 \subset K_2 \subset \cdots \cdots K_k^0 \subset K_{k+1} \subset \cdots$$

From the previous paragraph, $\{f_j\}$ has a subsequence which converges uniformly on K_1. Applying the same argument to this subsequence, we see that the subsequence has itself a subsequence which converges uniformly on K_2. Continuing in this manner, we construct an infinite matrix $\{f_{kj}\}$ of functions. The first row is the sequence $\{f_j\}$; each row is a subsequence of the previous row and, for each $k = 1, 2, \cdots$, the k-th row converges uniformly on K_k. The diagonal sequence $\{f_{kk}\}$ is thus a subsequence of $\{f_j\}$ which converges uniformly on each $K_m, m = 1, 2, \cdots$.

Now, let K be an arbitrary compact subset of Ω. Since $\{K_m^0\}$ is a nested open cover of Ω, it follows from compactness that K is contained in some K_m. Since $\{f_{kk}\}$ converges uniformly on K_m it also converges uniformly on K. □

Let Ω be an open set in $\mathbb{R}^n$. Denote by $C(\Omega)$ the family of continuous complex-valued functions on Ω. Fix an exhaustion $\{K_j\}$ of Ω and for $f, g \in C(\Omega)$, denote

$$d_j(f, g) = \sup_{z \in K_j} |f(z) - g(z)|$$

and

$$d(f, g) = \sum_{j=1}^{\infty} \frac{1}{2^j} \frac{d_j(f, g)}{1 + d_j(f, g)}.$$

Problem 20. Let Ω be an open subset of $\mathbb{R}^n$. Show that d is a distance function on $C(\Omega)$, that the induced metric space is complete and separable, and that a sequence of functions in $C(\Omega)$ converges with respect to this distance if and only if it converges uniformly on compact subsets of Ω. The induced topology on $C(\Omega)$ is called the topology of uniform convergence on compacta. Show that the space $C(\Omega)$ is a topological algebra.

Problem 21. Show that the space $\mathcal{O}(\Omega)$ of holomorphic functions on Ω is a closed subalgebra of $C(\Omega)$.

Chapter 5
Series

Abstract The basis of this chapter is Cauchy's theorem on multiple series. The Cauchy integral formula leads to the Taylor series representation for holomorphic functions, which in turn leads to the equivalence of the notions of analytic and holomorphic functions of several complex variables.

In the introduction, we asserted that holomorphic functions are the same as (complex) analytic functions. In order to discuss analytic functions of several variables, we must first discuss multiple series. We are inspired by the presentation in Range [13]. As with ordinary series, by abuse of notation, the expression

$$\sum_{\alpha \in \mathbb{N}^n} c_\alpha, \quad c_\alpha \in \mathbb{C}.$$

will have two meanings depending on the context. The first meaning is that this is simply a formal expression which we call a multiple series. The second meaning will be the sum of this multiple series, when it exists. Of course we now have to define what we mean by the sum of a multiple series. If $n > 1$, the index set $\mathbb{N}^n$ does not carry any natural ordering, so that there is no canonical way to consider $\sum c_\alpha$ as a sequence of (finite) partial sums as in the case $n = 1$. The ambiguity is avoided if one considers absolutely convergent series as follows. The multiple series $\sum_{\alpha \in \mathbb{N}^n} c_\alpha$ is called *absolutely convergent* if

$$\sum_{\alpha \in \mathbb{N}^n} |c_\alpha| = \sup \left\{ \sum_{\alpha \in F} |c_\alpha| : F \text{ finite} \right\} < \infty.$$

An absolutely convergent series is the same as an element of $L^1(\mathbb{N}^n, \mu)$, where μ is the counting measure.

Cauchy's theorem on multiple series asserts that *the absolute convergence of* $\sum c_\alpha$ *is necessary and sufficient for the following to hold.*
Any arrangement of $\sum c_\alpha$ *into an ordinary series*

$$\sum_{j=1}^{\infty} c_{\sigma(j)},$$

© Springer International Publishing Switzerland 2014
P.M. Gauthier, *Lectures on Several Complex Variables*,
DOI 10.1007/978-3-319-11511-5_5

where $\sigma : \mathbb{N} \to \mathbb{N}^n$ is a bijection, converges in the usual sense to a limit $L \in \mathbb{C}$ which is independent of σ. This number L is called the *limit* (or *sum*) of the multiple series, and one writes

$$\sum_{\alpha \in \mathbb{N}^n} c_\alpha = L.$$

It follows that, if $\sum c_\alpha$ converges absolutely, then its limit can be expressed as the sum of the *homogeneous* expansion

$$L = \sum_{k=1}^{\infty} \left(\sum_{|\alpha|=k} c_\alpha \right).$$

Moreover, if τ is a permutation of $\{1, \cdots, n\}$, then the iterated series

$$\sum_{\alpha_{\tau(n)}=1}^{\infty} \left(\cdots \left(\sum_{\alpha_{\tau(1)}=1}^{\infty} c_{\alpha_1 \cdots \alpha_n} \right) \cdots \right)$$

also converges to L. Here, as in any mathematical expression, we first perform the operation in the innermost parentheses and work our way out. Conversely, if $c_\alpha \geq 0$, and some iterated series converges, the convergence of $\sum c_\alpha$ follows.

The Cauchy theorem on multiple series follows immediately from the Fubini–Tonelli theorem, but we shall finesse integration theory and prove the Cauchy theorem for the case $n = 2$, that is, for double series.

Suppose a double series $\sum c_{jk}$ converges absolutely. Then, any arrangement of $\sum c_{jk}$ into a simple series converges absolutely. We know that if a simple series converges absolutely then it converges and any rearrangement converges to the same sum. Since any two arrangements of $\sum c_{jk}$ into simple series are rearrangements of each other, it follows that all arrangements of $\sum c_{jk}$ into simple series converge and to the same sum L. This proves the first part of Cauchy's double series theorem.

Now, let $\mathcal{P}^1, \mathcal{P}^2, \cdots$ be any partition of the set $\mathbb{N} \times \mathbb{N}$ of indices of the double series $\sum c_{jk}$. Cauchy's double series theorem further asserts that

$$L = \sum_{\nu} \left(\sum_{\mathcal{P}^\nu} c_{jk} \right).$$

We may consider each $\sum_{\mathcal{P}^\nu} c_{jk}$ as a double series obtained from the double series $\sum c_{jk}$ by possibly setting some of the terms equal to zero. Since the double series $\sum c_{jk}$ converges absolutely, it follows that the double series $\sum_{\mathcal{P}^\nu} c_{jk}$ also converges absolutely. Hence it converges. Denote the sum of $\sum_{\mathcal{P}^\nu} c_{jk}$ by $L_{\mathcal{P}^\nu}$. We must show that

$$L = \sum_{\nu} L_{\mathcal{P}^\nu}.$$

Fix $\epsilon > 0$. Let $\sum_{i=1}^{\infty} c_{\sigma(i)}$ be any arrangement of $\sum c_{jk}$ and choose n_1 so large that

$$\sum_{i=n_1}^{\infty} |c_{\sigma(i)}| < \epsilon.$$

Now choose n_2 so large that each of the terms $c_{\sigma(i)}, i < n_1$ is in one of the $\mathcal{P}^{\nu}, \nu < n_2$. Set $n(\epsilon) = \max\{n_1, n_2\}$. For $n > n(\epsilon)$ we have

$$|L - \sum_{\nu=1}^{n} L_{\mathcal{P}^{\nu}}| = |\sum_{i=1}^{\infty} c_{\sigma(i)} - \sum_{\nu=1}^{n} L_{\mathcal{P}^{\nu}}| <$$

$$|\sum_{i=n_1}^{\infty} c_{\sigma(i)} - \sum_{\nu=1}^{n} L'_{\mathcal{P}^{\nu}}| < \epsilon + \sum_{\nu=1}^{n} |L'_{\mathcal{P}^{\nu}}|,$$

where $L'_{\mathcal{P}^{\nu}}$ is the sum $L_{\mathcal{P}^{\nu}}$ from which those $c_{\sigma(i)}$ for which $i < n_1$ (if there are any such) have been removed. We note that

$$\sum_{\nu=1}^{n} |L'_{\mathcal{P}^{\nu}}| \le \lim_{m \to \infty} \sum_{\nu=1}^{n} \sum \{|c_{\sigma(i)}| : \sigma(i) \in \mathcal{P}^{\nu}, n_1 \le i < m\} \le \sum_{i=n_1}^{\infty} |c_{\sigma(i)}| < \epsilon.$$

Combining the above estimates, we have that, for $n > n(\epsilon)$,

$$|L - \sum_{\nu=1}^{n} L_{\mathcal{P}^{\nu}}| < 2\epsilon,$$

which concludes the proof of Cauchy's theorem for double series.

We recall the following from undergraduate analysis.

Theorem 20 (Weierstrass M-test). *Let f_n be sequence of functions defined on a set E and M_n a sequence of constants. If $|f_n| \le M_n$ and $\sum M_n$ converges, then $\sum f_n$ converges absolutely and uniformly.*

Problem 22. For $\zeta \in \mathbb{D}^n$, and recalling the abusive notation $1 = (1, \cdots, 1)$ as well as the notation given by (1) show that

$$\frac{1}{1 - \zeta} = \sum_{\alpha \ge 0} \zeta^{\alpha};$$

the series converges absolutely and any arrangement converges uniformly on compact subsets of $\mathbb{D}^n$.

The next theorem asserts that holomorphic functions are analytic. The converse will come later.

Theorem 21. *Let f be holomorphic in a domain $\Omega \subset \mathbb{C}^n$ and let $a \in \Omega$. Then, f can be expanded in an absolutely convergent power series:*

$$f(z) = \sum_{\alpha \geq 0} c_\alpha (z - a)^\alpha,$$

in a neighborhood of a. The series is the Taylor series of f; that is,

$$c_\alpha = \frac{f^{(\alpha)}(a)}{\alpha!}.$$

The representation of f as the sum of its Taylor series is valid in any polydisc centered at a and contained in Ω.

Proof. Consider a polydisc

$$\mathcal{D}(a, r) = \{z : |z_j - a_j| < r, j = 1, \cdots, n\},$$

whose closure is contained in Ω. By the Cauchy integral formula,

$$f(z) = \frac{1}{(2\pi i)^n} \int_{b\mathcal{D}} \frac{f(\zeta)}{\zeta - z} d\zeta.$$

By an earlier problem, we may write

$$\frac{f(\zeta)}{\zeta - z} = \frac{f(\zeta)}{(\zeta - a) - (z - a)} = \frac{f(\zeta)}{\zeta - a} \cdot \frac{1}{1 - \frac{z-a}{\zeta-a}} = \frac{f(\zeta)}{\zeta - a} \sum_{\alpha \geq 0} \left(\frac{z - a}{\zeta - a} \right)^\alpha$$

and the convergence is uniform on $b\mathcal{D} \times K$ for any compact subset K of $\mathcal{D}$. Integrating term by term, we have

$$f(z) = \sum_{\alpha \geq 0} \left(\frac{1}{(2\pi i)^n} \int_{b\mathcal{D}} \frac{f(\zeta)}{(\zeta - a)^{\alpha+1}} d\zeta \right) (z - a)^\alpha = \sum_{\alpha \geq 0} c_\alpha (z - a)^\alpha$$

and the convergence is uniform on compact subsets of $\mathcal{D}$. By the Cauchy formula for derivatives, $c_\alpha = f^{(\alpha)}(a)/\alpha!$.

We have assumed that the closure of the polydisc is contained in Ω, but any polydisc contained in Ω can be written as the union of an increasing sequence of polydiscs with the same center whose closures are contained in Ω. The function f is represented by its Taylor series about a for each of the polydiscs in this sequence and hence the representation is valid on the union of these polydiscs. □

The previous proof extends to the case where the radii of the polydisc are not necessarily equal. From this we have interesting consequences if Ω is a polydisc or ball centered at a point a.

Since a polydisc $\mathcal{D}$ centered at a point a is the union of an increasing sequence of closed polydiscs centered at a, it follows that a function holomorphic in $\mathcal{D}$ is represented uniformly on compact subsets by its Taylor series about the point a.

If $\mathcal{B}$ is a ball centered at a point a, then every point of $\mathcal{B}$ is contained in a closed polydisc centered at the point a and contained in $\mathcal{B}$. It follows that a function holomorphic in a ball is represented locally uniformly by its Taylor series about the center of the ball.

We have now established that holomorphic functions are analytic. In the proof we did not require the property that holomorphic functions are C^1. We merely required uniform convergence to allow us to integrate term by term, and for this it is sufficient that holomorphic functions be locally bounded.

To prove conversely that analytic functions are holomorphic, we need a little more familiarity with multiple power series.

Theorem 22 (Abel). *If the power series $\sum c_\alpha z^\alpha$ converges at the point a for some arrangement (as a simple series) and if $a_j \neq 0, j = 1, \cdots, n$, then the series converges absolutely and uniformly on each compact subset of the polydisc*

$$\{z : |z_j| < |a_j|, j = 1, \cdots, n\}.$$

Proof. Since some arrangement of the series converges, it follows that the terms are bounded. Thus $|c_\alpha a^\alpha| < M$ for all α. Fix $0 < r_j < |a_j|, j = 1, \cdots, n$ and suppose $|z_j| \leq r_j, j = 1, \cdots, n$. Then,

$$|c_\alpha z^\alpha| = |c_\alpha z_1^{\alpha_1} \cdots z_n^{\alpha_n}| = |c_\alpha a_1^{\alpha_1} \cdots a_n^{\alpha_n}| \cdot \left| \left(\frac{z_1}{a_1} \right)^{\alpha_1} \cdots \left(\frac{z_n}{a_n} \right)^{\alpha_n} \right| \leq$$

$$M \left| \frac{r_1}{a_1} \right|^{\alpha_1} \cdots \left| \frac{r_n}{a_n} \right|^{\alpha_n} = M\rho^\alpha,$$

where $\rho_j < 1, j = 1, \cdots, n$. Since $\sum \rho^\alpha$ converges, the power series converges absolutely and uniformly on the closed polydisc $|z_j| \leq r_j, j = 1, \cdots, n$, by the Weierstrass M-test. Since any compact subset of the open polydisc $\{z : |z_j| < |a_j|, j = 1, \cdots, n\}$ is contained in such a closed polydisc, the proof is complete. $\square$

Let Σ_α be a multiple power series. Abel's theorem asserts that Σ_α converges absolutely and uniformly on compact subsets of the polydisc $|z_j| < |a_j|, j = 1, \cdots, n$, if some arrangement of Σ_α converges at the point a. Suppose we write $a = (b, c)$ and Σ_α as the iteration $\Sigma_\beta \Sigma_\gamma$ of two power series, which in some sense converges at (b, c). Can we hope for the same conclusion that Σ_α converges absolutely and uniformly on compact subsets of the polydisc $|z_j| < |a_j|, j = 1, \cdots, n$? Our meaning will be made clear by the following example which illustrates the futility of such a hope.

Example. Write $z = (\zeta, w)$ and consider the double power series

$$\sum_\alpha z^\alpha = \sum_{j,k} c_{j,k} \zeta^j w^k,$$

where $c_{j,j} = 4^j, c_{j,j+1} = -4^j$, and $c_{j,k} = 0$ if k is different from j or $j + 1$. Then, for $z = (\zeta, w) = (1, 1)$,

$$\sum_j c_{j,k} 1^j = 4^j - 4^j = 0, \quad k = 0, 1, \cdots,$$

and consequently,

$$\sum_k \left(\sum_j c_{j,k} 1^j \right) 1^k = \sum_k 0 \cdot 1^k = 0.$$

It is certainly not true that the double power series converges absolutely on the polydisc $|\zeta| < 1, |w| < 1$. This would imply that for any such point (ζ, w), the terms $c_{j,k} \zeta^j w^k$ would tend to zero. However, for the point $(1/2, 1/2)$, the "diagonal" terms are

$$c_{j,j} \left(\frac{1}{2}\right)^j \left(\frac{1}{2}\right)^j = 4^j \left(\frac{1}{2}\right)^{2j} = 1.$$

Theorem 23. *On an open set Ω in $\mathbb{C}^n$, a function is holomorphic if and only if it is analytic.*

Proof. We have shown earlier that every holomorphic function is analytic. Conversely, suppose f is analytic on Ω. It is sufficient to show that f is holomorphic in a polydisc about each point of Ω. Fix $a \in \Omega$ and let $\mathcal{D}$ be a polydisc containing a and contained in Ω, such that f can be represented as a power series in $\mathcal{D}$. We have seen that the power series converges uniformly on compact subsets of $\mathcal{D}$. In particular, let Q be a polydisc containing a and whose closure is compact in $\mathcal{D}$. Then the power series converges uniformly in Q and, since the terms are polynomials, they are holomorphic. Thus, f is the uniform limit of holomorphic functions on Q. Hence, f is holomorphic on Q. We have shown that f is holomorphic in a neighborhood of each point of Ω and so f is holomorphic in Ω. $\square$

Chapter 6
Zero Sets of Holomorphic Functions

Abstract Zeros of holomorphic functions of several variables have a much richer structure than those of a single variable. Fundamental concepts from algebra enter the scene and lead up to the Weierstrass preparation theorem, which is the best instrument for understanding the local nature of the set of zeros of a holomorphic function.

As in one complex variable, zero sets of holomorphic functions are very important in several complex variables. For example, consider the holomorphic function of two complex variables $f(z, w) = zw$. The zero set is the union of the z-axis and the w-axis, which are both complex lines. Thus, it is a complex line in the neighborhood of every one of its points except the origin. This example is rather typical in the sense that the zero set of a holomorphic function is very nice in the neighborhood of most of its points.

Lemma 24. *If f is holomorphic in a domain $\Omega \subset \mathbb{C}^n$ and not identically zero, then the zero set $Z(f)$ of f is a closed nowhere dense set in Ω.*

Proof. That $Z(f)$ is closed follows from the continuity of f. That $Z(f)$ is nowhere dense follows from the uniqueness Theorem 1. □

If f is a function holomorphic in a neighborhood of $0 \in \mathbb{C}^n$, we can write it as a power series in the z_n-variable, whose coefficients are holomorphic functions of $z_1, \cdots, z_{n-1}$, by writing the power series for f as an iterated series:

$$f(z) = \sum_\alpha a_\alpha z^\alpha = \sum_{\alpha_n} \cdots \sum_{\alpha_1} a_\alpha z_1^{\alpha_1} \cdots z_n^{\alpha_n} = \sum_{j=0}^{\infty} f_j(z_1, \cdots, z_{n-1}) z_n^j.$$

A function f holomorphic in a neighborhood of $0 \in \mathbb{C}^n$ is said to be *of order k in the z_n variable at* 0, if $f_j(0) = 0$, for $0 < j < k$ and $f_k(0) \neq 0$. If f is holomorphic in a neighborhood of a point $p \in \mathbb{C}^n$, we say f is *of order k in the z_n variable at p*, if the function $z \mapsto f(z + p)$ is of order k in the z_n variable at 0.

Lemma 25. *If f is holomorphic and nonconstant in a neighborhood of a point $p \in \mathbb{C}^n$, then, after a linear change of variables, f is of order k in the z_n variable at p, for some $k \geq 1$.*

© Springer International Publishing Switzerland 2014
P.M. Gauthier, *Lectures on Several Complex Variables*,
DOI 10.1007/978-3-319-11511-5_6

Proof. Consider the power series expansion of f in a neighborhood of p.

$$f(z) = f(p) + \sum_{|\alpha| \geq 1} a_\alpha (z - p)^\alpha, \quad |z - p| < \delta.$$

Since, f is not constant, $a_\alpha \neq 0$, for some $\alpha = \alpha_1 + \cdots + \alpha_n$, with $|\alpha| \geq 1$. Thus, $\alpha_j \geq 1$, for some j. By changing coordinates, we may assume that $j = n$. In these coordinates, if we write the power series as a power series in the z_n-variable, with coefficients holomorphic in $z_1, \cdots, z_{n-1}$, we have

$$f(z) = f(p) + \sum_{j=1}^{\infty} f_j(z_1, \cdots, z_{n-1})(z_n - p_n)^j,$$

where not all the $f_j(p_1, \cdots, p_{n-1})$ vanish. If k is the smallest such j, then f is of order k in the z_n-direction at p.   □

Let us write

$$z = (z_1, \cdots, z_n) = (z', z_n), \quad z' \in \mathbb{C}^{n-1}.$$

If f is holomorphic in a neighborhood of 0, we may write

$$f(z) = \sum_{j=k}^{\infty} f_j(z') z_n^j = f_{z'}(z_n), \quad |z'| < r, |z_n| < r, \tag{6.1}$$

where $|z'|$ is the Euclidean norm and where each $f_{z'}$ is a holomorphic function of z_n in $|z_n| < r$.

Theorem 26. *For $n > 1$, if f is holomorphic in a neighborhood of 0 and has a zero of order $k \geq 1$ in the z_n direction, then, for all $r > 0$, and all sufficiently small $z' \in \mathbb{C}^{n-1}$, $f_{z'}(z_n) = f(z', z_n)$ has at least k zeros (counting multiplicities) in $|z_n| < r$. In particular, a nonconstant holomorphic function in a domain of $\mathbb{C}^n, n > 1$, has no isolated zeros.*

Proof. If f is not identically zero, then, by the preceding lemma, we may assume that f is of order $k \geq 1$ in the z_n direction at 0. We have, by (6.1),

$$f(z) = \sum_{j=k}^{\infty} f_j(z') z_n^j = f_{z'}(z_n), \quad |z'| < r, |z_n| < r,$$

where each $f_{z'}$ is a holomorphic function of z_n in $|z_n| < r$. Since $f_{0'}$ has a zero of order $k \geq 1$ at 0, there is an $r_o < r$, such that $f_{0'}(z_n) \neq 0$, for $0 < |z_n| \leq r_o$. Fix ϵ, with $0 < \epsilon < \min |f_{0'}(z_n)|$, for $|z_n| = r_o$. For sufficiently small $\delta > 0$, and $|z'| < \delta$, we have

$$\max_{|z_n| = r_o} |f_{z'}(z_n) - f_{0'}(z_n)| < \epsilon.$$

Since $f_{0'}$ has a zero of order $k \geq 1$ at 0, it follows from Rouché's theorem that for $|z'| < \delta$, the functions $f_{z'}$ have at least $k \geq 1$ zeros in $|z_n| < r_o$. In other words, for all small z', and hence for all small $z' \neq 0$, $f(z', z_n)$ has a zero, with $|z_n| < r_o$. We may choose r arbitrarily small and repeat this process. Thus, f has zeros different from 0 in each neighborhood of 0.   □

The fundamental theorem of algebra asserts that every nonconstant complex polynomial of a single variable has at least one zero. An extremely important consequence is that every such polynomial $p(z)$ has a factorization

$$p(z) = \alpha(z - \alpha_1) \cdots (z - \alpha_k),$$

where α is a nonzero complex number and the α_j are the zeros of p, possibly repeated according to multiplicities. Moreover, such a factorization is unique up to order.

We wish to consider polynomials of several variables. Let us recall a few facts from algebra. A unique factorization domain (UFD) is an integral domain R with the property that every nonzero element x has a factorization $x = ux_1 \cdots x_m$, where the x_j are irreducible and u is a unit. This factorization is unique up to changing the order of the x_j or multiplying the x_j by units. If R is a UFD, then so is the ring $R[X]$ of polynomials. Invoking this fact, the ring $R[X_1, X_2]$ of polynomials in two variables, with coefficients in R, is a UFD, since $R[X_1, X_2] = (R[X_1])[X_2]$. Repeating finitely many times, we have that, if R is a UFD, then the ring $R[X_1, \cdots, X_n]$ of polynomials in n variables with coefficients in R is also a UFD. In particular, for an arbitrary field F, the ring $F[X_1, \cdots, X_n]$ is a UFD.

The polynomials we have been discussing are not the usual polynomial functions but rather *formal* polynomials in indeterminates $X_1, \cdots, X_n$. But since we wish to do analysis more than algebra, we are more interested in the usual polynomial *functions*. Fortunately, if the ring R is infinite, then the ring $R[X]$ of polynomials $p(X) = \sum a_j X^j$ in the indeterminate X, with coefficients a_j in R, is isomorphic to the ring $R[x]$ of polynomial functions $p : R \to R$, defined, for $x \in R$, by $p(x) = \sum_j a_j x^j$. Applying the above considerations to the rings $\mathbb{C}, \mathbb{C}[z], \mathbb{C}[z_1, z_2]$, etc., we arrive at the following.

Theorem 27. *The ring* $\mathbb{C}[z_1, \cdots, x_n]$ *of polynomials (as functions) of n complex variables is a unique factorization domain.*

It can be shown that, since $\mathbb{C}[z_1, \cdots, x_n]$ is a UFD, any two polynomials $p, q \in \mathbb{C}[z_1, \cdots, x_n]$ have a greatest common divisor (gcd). Since the polynomials are not linearly ordered, we must explain what is meant by gcd. A polynomial ϕ is a gcd of p and q if ϕ divides both p and q and, if ψ is another polynomial which divides p and q, then ψ divides ϕ. A gcd is not unique, but almost. It is unique up to multiplication by a nonzero complex number.

The Weierstrass preparation theorem is an important step in extending our knowledge on factorizing polynomials towards factorizing holomorphic functions in an interesting and useful way. It is the most fundamental result giving a local description of such zero sets.

Theorem 28 (Weierstrass preparation theorem). *Let f be holomorphic in a polydisc D centered at the origin in $\mathbb{C}^n$. Suppose $f(0,\cdots,0) = 0$, but $f(0,\cdots, 0,z_n) \not\equiv 0$. Then, there is a neighborhood of the origin in which f can be represented as follows:*

$$f = g \cdot (z_n^m + p_1 z_n^{m-1} + \cdots + p_m),$$

where g is holomorphic and zero free, while p_j are holomorphic functions in $z_1,\cdots,z_{n-1}$ which vanish at the origin in $\mathbb{C}^{n-1}$, $j = 1,\cdots,m$.

Proof. For simplicity, we prove the theorem for $n = 2$. Since $f(0,z_2) \not\equiv 0$, we can choose small $r > 0$, such that $f(0,z_2) \neq 0$, for $|z_2| = r$. By continuity, we may choose a small $\delta > 0$ such that $f(z_1,z_2) \neq 0$, for $|z_2| = r$ and $|z_1| < \delta$. For each z_1 with $|z_1| < \delta$, the number $N(z_1)$ of zeros (counting multiplicities) of $f(z_1,\cdot)$ in $|z_2| < r$ is given by the argument principle

$$N(z_1) = \frac{1}{2\pi i} \int_{|z_2|=r} \frac{(\partial f/\partial z_2)(z_1,z_2)}{f(z_1,z_2)} dz_2.$$

Notice that the right side of this equation is a continuous function of z_1, while the left side is an integer-valued function. Thus, $N(z_1)$ is constant and so $N(z_1) = N(0) = m$, the order of the zero of $f(0,\cdot)$ at the origin. For $|z_1| < \delta$, let $\eta_1(z_1),\cdots,\eta_m(z_1)$ be the zeros of $f(z_1,\cdot)$ in $|z_2| < r$. Set

$$F(z_1,z_2) = (z_2 - \eta_1(z_1))(z_2 - \eta_2(z_1))\cdots(z_2 - \eta_m(z_1)) =$$

$$z_2^m + p_1(z_1)z_2^{m-1} + \cdots + p_m(z_1).$$

For each z_1, $F(z_1,\cdot)$ is the monic polynomial whose zeros coincide with those of $f(z_1,\cdot)$. Putting $z_1 = 0$, we see that $p_j(0) = 0$, $j = 1,\cdots,m$. We shall show that we may set $g = f/F$.

It is known from algebra that, for fixed z_1, the coefficients $p_j(z_1)$ can be expressed as polynomials in the symmetric functions

$$S_k(z_1) = \eta_1^k + \cdots \eta_m^k, \quad k = 1,\cdots,m.$$

Hence, to show that the p_j are holomorphic, it is sufficient to show that the symmetric functions $S_k(z_1)$ are holomorphic. By residue theory,

$$S_k(z_1) = \frac{1}{2\pi i} \int_{|z_2|=r} z_2^k \frac{(\partial f/\partial z_2)(z_1,z_2)}{f(z_1,z_2)} dz_2$$

and so S_k is given by an expression

$$S_k(z_1) = \int_C \phi(z_1,z_2)dz_2, \quad |z_1| < \delta,$$

where, on $(|z_1| < \delta) \times (|z_2| = r)$, the function $\phi(z_1, z_2)$ is continuous and holomorphic in z_1. Thus, by Leibniz's theorem, S_k is holomorphic in $|z_1| < \delta$. Thus, the function F is holomorphic in $(|z_1| < \delta) \times \mathbb{C}$.

For each $|z_1| < \delta$, we set $g(z_1, \cdot) = f(z_1, \cdot)/F(z_1, \cdot)$, which is a well-defined holomorphic function of z_2, having no zeros for $|z_2| < r$. By the Cauchy Formula, for each $|z_1| < \delta$,

$$g(z_1, z_2) = \frac{1}{2\pi i} \int_{|\zeta_2|=r} \frac{g(z_1, \zeta_2)}{\zeta_2 - z_2} d\zeta_2. \tag{6.2}$$

Now, F is holomorphic on $(|z_1| < \delta) \times \mathbb{C}$ and for each fixed z_1 has all of its zeros in $|z_2| < r$. Thus F has no zeros on $(|z_1| < \delta) \times (|\zeta_2| = r)$. Hence, $1/F$ is holomorphic on $(|z_1| < \delta) \times (|\zeta_2| = r)$ and so is g. By the Leibniz rule, one can partially derive the integral on the right side of (6.2), by differentiating under the integral sign and conclude that the right side is holomorphic in $(|z_1| < \delta) \times (|\zeta_2| < r)$. Thus, the left side g is also holomorphic there.   □

Chapter 7
Holomorphic Mappings

Abstract The theory of smooth mappings of several real variables will help us to develop a similar theory for smooth (holomorphic) mappings of several complex variables. As in the real case, the main features are the inverse function theorem and the implicit function theorem.

In this section holomorphic mappings are studied by translating the real setting into the complex setup (implicit function theorem, rank theorem). For an excellent and more thorough presentation, we refer to [8].

Problem 23 (Chain rule). Suppose $\zeta \to z$ is a smooth mapping from an open set $D \subset \mathbb{C}$ into an open set $\Omega \subset \mathbb{C}^n$ and $z \to w$ is a smooth function from Ω into $\mathbb{C}$, then

$$\frac{\partial w}{\partial \zeta} = \sum_{j=1}^{n} \left(\frac{\partial w}{\partial z_j} \frac{\partial z_j}{\partial \zeta} + \frac{\partial w}{\partial \bar{z}_j} \frac{\partial \bar{z}_j}{\partial \zeta} \right)$$

and

$$\frac{\partial w}{\partial \bar{\zeta}} = \sum_{j=1}^{n} \left(\frac{\partial w}{\partial z_j} \frac{\partial z_j}{\partial \bar{\zeta}} + \frac{\partial w}{\partial \bar{z}_j} \frac{\partial \bar{z}_j}{\partial \bar{\zeta}} \right).$$

A mapping $f : \Omega \to \mathbb{C}^m$, defined on an open subset Ω of $\mathbb{C}^n$, is said to be *holomorphic* if each of the components $f_1, \cdots, f_m$ of f is holomorphic.

Problem 24. The composition of holomorphic mappings is holomorphic. That is, if D is an open subset of $\mathbb{C}^k$, Ω is an open subset of $\mathbb{C}^n$, $g : D \to \mathbb{C}^n$ and $f : \Omega \to \mathbb{C}^m$ are holomorphic mappings, and $g(D) \subset \Omega$, then the mapping $f \circ g : D \to \mathbb{C}^m$ is holomorphic.

Let $f : \Omega \to \mathbb{C}^m$ be a holomorphic mapping defined in an open set $\Omega \subset \mathbb{C}^n$. To each $z \in \Omega$, we associate a unique $\mathbb{C}$-linear transformation $f'(z) : \mathbb{C}^n \to \mathbb{C}^m$, called the *derivative of f at z*, such that

$$f(z + h) = f(z) + f'(z)h + r(h),$$

where $r(h) = O(|h|^2)$ as $h \to 0$.

© Springer International Publishing Switzerland 2014

P.M. Gauthier, *Lectures on Several Complex Variables*,

DOI 10.1007/978-3-319-11511-5_7

Problem 25. Prove the uniqueness of the derivative.

With respect to the standard coordinates in $\mathbb{C}^n$ and $\mathbb{C}^m$, the linear transformation $f'(a)$ at a point a is represented by the (complex) Jacobian matrix

$$J(f)(a) = \left(\frac{\partial f_j}{\partial z_k}(a)\right), \quad j = 1, \cdots, m; \, k = 1, \cdots, n.$$

Of course this matrix represents a linear transformation, so we need only verify that it has the required approximation property. Since the vector $r(h)$ is small if and only if each of its components is small, it is sufficient to check the claim for each component f_j of f. Thus, it is enough to suppose that f itself is a function rather than a mapping. From the Taylor formula,

$$f(z+h) = f(z) + J(f)(z)h + \sum_{|\alpha| \geq 2} \frac{f^{(\alpha)}(z)}{\alpha!} h^\alpha.$$

Write $h = t\zeta$, with t positive. Then,

$$r(h) = \sum_{k=2}^\infty \left(\sum_{|\alpha|=k} \frac{f^{(\alpha)}(z)}{\alpha!}\zeta^\alpha\right) t^k = t^2 \sum_{k=0}^\infty \left(\sum_{|\alpha|=k+2} \frac{f^{(\alpha)}(z)}{\alpha!}\zeta^\alpha\right) t^k.$$

Since the original power series in h converges absolutely for small h, the power series in ζ and t converges for some positive t and some ζ, none of whose coordinates are zero. It follows that the series

$$\sum_{k=0}^\infty \left(\sum_{|\alpha|=k+2} \frac{f^{(\alpha)}(z)}{\alpha!}\zeta^\alpha\right) t^k$$

in ζ and t converges for all small ζ and t. Thus, for some $t_0 > 0$ and $\rho > 0$, this sum is bounded, by say M, for $|t| \leq t_0$ and $|\zeta| \leq \rho$. If $|h| \leq t_0\rho$, we may write

$$h = \left(\frac{|h|}{\rho}\right)\left(\frac{h}{|h|}\rho\right) = t\zeta,$$

with $|t| \leq t_0$ and $|\zeta| \leq \rho$. Thus, for $|h| \leq t_0\rho$, we have

$$|r(h)| \leq t^2 M = \left(\frac{|h|}{\rho}\right)^2 M = O(|h|^2),$$

which concludes the proof that f' is represented by the Jacobian matrix $J(f)$.

Problem 26. If f and g are holomorphic mappings such that $f \circ g$ is defined, then $(f \circ g)' = (f'(g))g'$ and $J(f \circ g) = J(f)J(g)$. More precisely, if $w = g(z)$, then $(f \circ g)'(z) = f'(w)g'(z)$ and $J(f \circ g)(z) = [J(f)(w)][J(g)(z)]$.

Let Ω be open in $\mathbb{C}^n$ and $f : \Omega \to \mathbb{C}^m$ be a mapping, which we may write $f(z) = w$, with $z \in \Omega$ and $w \in \mathbb{C}^m$. Let x and y be the real and imaginary parts of z and let u and v be the real and imaginary parts of w. We may think of Ω as an open subset of $\mathbb{R}^{2n}$ and we may view the complex mapping $z \mapsto w$ as a real mapping $(x, y) \mapsto (u, v)$ of the open subset Ω of $\mathbb{R}^{2n}$ into $\mathbb{R}^{2m}$. If the complex mapping f is smooth, let $J_{\mathbb{R}}(f)$ denote the (real) Jacobian matrix of the associated (real) mapping $(x, y) \mapsto (u, v)$. If f is an equidimensional smooth complex mapping, then the complex and real Jacobian matrices $J(f)$ and $J_{\mathbb{R}}(f)$ are square. $\det J(f)$ is called the (complex) *Jacobian determinant* of f and $\det J_{\mathbb{R}}(f)$ is called the *real Jacobian determinant* of f.

Theorem 29. *If f is an equidimensional holomorphic mapping, then*

$$\det J_{\mathbb{R}}(f) = |\det J(f)|^2.$$

Problem 27. Verify this for $n = 1$.

Proof. In this proof we shall sometimes denote the determinant of a square matrix A by $|A|$. We shall write matrices as block matrices, where for example $\partial u / \partial x$ represents the matrix $(\partial u_j / \partial x_k)$. Since an even number of permutations of rows and columns do not change the determinant, we may write

$$\det J_{\mathbb{R}}(f) = \begin{vmatrix} \frac{\partial u}{\partial x} & \frac{\partial u}{\partial y} \\ \frac{\partial v}{\partial x} & \frac{\partial v}{\partial y} \end{vmatrix}.$$

Adding a constant multiple of a row to another row does not change the determinant, so we may add i times the lower blocks to the upper blocks and use the Cauchy–Riemann equations to obtain

$$\det J_{\mathbb{R}}(f) = \begin{vmatrix} \frac{\partial u}{\partial x} + i \frac{\partial v}{\partial x} & i \frac{\partial u}{\partial x} - \frac{\partial v}{\partial y} \\ \frac{\partial v}{\partial x} & \frac{\partial u}{\partial x} \end{vmatrix}.$$

Now, if we subtract i times the left blocks from the right blocks, we have

$$\det J_{\mathbb{R}}(f) = \begin{vmatrix} \frac{\partial f}{\partial x} & 0 \\ \cdots & \frac{\partial \overline{f}}{\partial x} \end{vmatrix}.$$

We have $\partial \overline{f} / \partial x = \overline{\partial f / \partial x}$ and since f is holomorphic $\partial f / \partial x = \partial f / \partial z$. Thus,

$$\det J_{\mathbb{R}}(f) = \begin{vmatrix} \frac{\partial f}{\partial z} & 0 \\ \cdots & \frac{\partial \overline{f}}{\partial z} \end{vmatrix} = \left| \frac{\partial f}{\partial z} \right| \left| \overline{\frac{\partial f}{\partial z}} \right| = \left| \frac{\partial f}{\partial z} \right| \overline{\left| \frac{\partial f}{\partial z} \right|} = \det J(f) \cdot \overline{\det J(f)}.$$

Hence,

$$\det J_{\mathbb{R}}(f) = |\det J(f)|^2.$$

$\square$

Theorem 30 (Inverse mapping). *Let f be a holomorphic mapping defined in a neighborhood of a point a. Then, f is invertible in a neighborhood of a and the inverse mapping is also holomorphic if and only if $f'(a)$ is invertible.*

Proof. Since f is holomorphic it is smooth. If $f'(a)$ is invertible it is equidimensional and so the Jacobian matrix $J(f)(a)$ is square and invertible. Thus, $\det J(f)(a) \neq 0$. By the previous theorem, $\det J_{\mathbb{R}}(f)(a) \neq 0$. Thus, we may invoke the real inverse mapping theorem to conclude that $w = f(z)$ considered as a real mapping is locally invertible at a. Let g denote the local inverse mapping defined in a neighborhood of $b = f(a)$. Then, g is smooth and, since $z = (g \circ f)(z)$, we have for $j = 1, \cdots, n$ and $k = 1, \cdots, n$:

$$0 = \frac{\partial z_j}{\partial \bar{z}_k} = \sum_\nu \frac{\partial g_j}{\partial w_\nu} \frac{\partial f_\nu}{\partial \bar{z}_k} + \sum_\nu \frac{\partial g_j}{\partial \bar{w}_\nu} \frac{\partial \bar{f}_\nu}{\partial \bar{z}_k} = \sum_\nu \frac{\partial g_j}{\partial \bar{w}_\nu} \frac{\partial \bar{f}_\nu}{\partial \bar{z}_k}.$$

Since

$$\frac{\partial \bar{f}_\nu}{\partial \bar{z}_k} = \overline{\frac{\partial f_\nu}{\partial z_k}},$$

we have the matrix equation

$$(0) = \left(\frac{\partial g}{\partial \bar{w}}\right) \overline{\left(\frac{\partial f}{\partial z}\right)} = \left(\frac{\partial g}{\partial \bar{w}}\right) \overline{J(f)}. \tag{7.1}$$

Now, since $f'(a)$ is invertible, $\det J(f)(a) \neq 0$ and so $\det J(f)(z) \neq 0$ for z in a neighborhood of a. Thus, $J(f)$ and consequently $\overline{J(f)}$ also is invertible for z in a neighborhood of a. Multiplying equation (7.1) on the right by the inverse matrix of $\overline{J(f)}$, we have

$$(0) = \left(\frac{\partial g}{\partial \bar{w}}\right).$$

That is, the components of g satisfy the Cauchy–Riemann equations in a neighborhood of b. Therefore, the inverse mapping g is also holomorphic in a neighborhood of $b = f(a)$.

Suppose, conversely, that f is locally invertible in a neighborhood of a and the inverse g is also holomorphic in a neighborhood of $b = f(a)$. Then, denoting the identity matrix I, we have $I = I'(a) = (g \circ f)'(a) = g'(f(a))f'(a)$. Therefore, $f'(a)$ is invertible.

$\square$

Corollary 31. *Let $f : \Omega \to \mathbb{C}^n$ be an equidimensional holomorphic mapping on a domain $\Omega \subset \mathbb{C}^n$. Then, f is locally biholomorphic if and only if $\det J_f(z) \neq 0$, for each $z \in \Omega$.*

Next, we shall present the *implicit mapping theorem*. But first, we shall try to motivate the formulation by an informal heuristic discussion, which the student should not take too seriously. Suppose f is a holomorphic mapping from an open set W in $\mathbb{C}^{n+m}$ to $\mathbb{C}^k$ and we would like the level set $f^{-1}(0)$ near a point (a, b) in $\mathbb{C}^{n+m}$ where $f(a, b) = 0$ to look like a graph in $\mathbb{C}^n \times \mathbb{C}^m$ of a function $w = g(z)$ defined in a neighborhood of a and taking its values in $\mathbb{C}^m$. First of all, we had better have $k = m$. Secondly, if we want the level set to be a graph over a neighborhood of a we would not wish the level set to be "vertical" at (a, b). In the real case with $n = m = 1$, we preclude this by the condition $\partial f / \partial y \neq 0$ at (a, b). In the multivariable situation, we preclude this strongly by asking that the matrix $\partial f / \partial y$ be invertible at (a, b).

Theorem 32 (Implicit mapping). *Let $f(z, w)$ be a holomorphic mapping from a neighborhood of a point (a, b) in $\mathbb{C}^{n+m}$ to $\mathbb{C}^m$ and suppose $f(a, b) = 0$. If*

$$\det \frac{\partial f}{\partial w}(a, b) \neq 0, \tag{7.2}$$

then there are neighborhoods U and V of a and b respectively and a holomorphic mapping $g : U \to V$ such that $f(z, w) = 0$ in $U \times V$ if and only if $w = g(z)$.

Proof. As in the proof of the inverse mapping theorem, we obtain all of the conclusions from the real implicit mapping theorem except the holomorphy of g. For $z \in U$, we have $f(z, g(z)) = 0$ and hence, for $j = 1, \cdots, m$; $k = 1, \cdots, n$:

$$0 = \frac{\partial f_j}{\partial \overline{z}_k} = \sum_\nu \frac{\partial f_j}{\partial z_\nu} \frac{\partial z_\nu}{\partial \overline{z}_k} + \sum_\nu \frac{\partial f_j}{\partial w_\nu} \frac{\partial g_\nu}{\partial \overline{z}_k} = \sum_\nu \frac{\partial f_j}{\partial w_\nu} \frac{\partial g_\nu}{\partial \overline{z}_k}.$$

Fix $z \in U$ and define $f_z(w) = f(z, w)$ for $w \in V$. Then, the preceding equations can be written as the matrix equation

$$(0) = \left(\frac{\partial f_z}{\partial w} \right) \left(\frac{\partial g}{\partial \overline{z}} \right). \tag{7.3}$$

By continuity, we may assume that (7.2) holds not only at (a, b) but also at all points $(z, w) \in U \times V$. Thus, for all $z \in U$, the Jacobian matrix $\partial f_z / \partial w$ is invertible at all points $w \in V$. Now, if we multiply both members of (7.3) on the left by the inverse of the matrix $\partial f_z / \partial w$, we obtain that the matrix $\partial g / \partial \overline{z}$ is the zero matrix. Thus, g satisfies the Cauchy–Riemann equations and is therefore holomorphic. □

The following version of the rank theorem follows the formulation in [8]. In the rank theorem, we assume we are given a holomorphic mapping f such that the rank of f' is a constant r near a point a. The simplest example would be when f is a linear transformation of rank r, for then $f'(x) = f$, for each x. The simplest example of a linear transformation of rank r from $\mathbb{C}^n$ to $\mathbb{C}^m$ is the mapping $(z_1, \cdots, z_n) \mapsto (z_1, \cdots, z_r, 0, \cdots, 0)$. The rank theorem asserts that near a the mapping f can be put in this form by a biholomorphic change of coordinates.

Theorem 33 (Rank). *Let f be a holomorphic mapping from a neighborhood of a point a in $\mathbb{C}^n$ to $\mathbb{C}^m$ and suppose f' has constant rank r near a. Then, there are neighborhoods U and V of a and $b = f(a)$, polydiscs $D^n \subset \mathbb{C}^n$ and $D^m \subset \mathbb{C}^m$, each centered at 0, and biholomorphic mappings $\varphi : D^n \to U$ and $\psi : V \to D^m$ with $\varphi(0) = a$ and $\psi(b) = 0$ such that, with $\chi(z_1, \cdots, z_n) = (z_1, \cdots, z_r, 0, \cdots, 0)$, we have $\chi = \psi \circ f \circ \varphi$.*

Proof. We may suppose that $a = b = 0$; moreover, we may choose the coordinates of $\mathbb{C}^n$ and $\mathbb{C}^m$ so that

$$f'(0) = \begin{pmatrix} I_r & 0 \\ 0 & 0 \end{pmatrix}.$$

Consider the mapping

$$g(z) = (f_1(z), \cdots, f_r(z), z_{r+1}, \cdots, z_n).$$

Clearly, we have $g'(0) = I_n$. By the inverse mapping theorem, there is an open neighborhood U of 0 in $\mathbb{C}^n$ such that g maps U biholomorphically onto a polydisc D^n. Set $\varphi := (g|_U)^{-1}$. It follows that for $w \in D^n$ and $z := \varphi(w)$, we have

$$(f \circ \varphi)_j(w) = f_j(\varphi(w)) = f_j(z) = w_j, \quad j = 1, \cdots, r.$$

Hence,

$$(f_1, \cdots, f_m)(z) = f(z) = (f \circ \varphi)(w) =: (w_1, \cdots, w_r, h_{r+1}(w), \cdots, h_m(w)),$$

where each h_j is holomorphic. For the mapping $f \circ \varphi$ we have that rank $(f \circ \varphi)' \geq r$ on D^n. By hypothesis rank $f' = r$ on D^n; thus, by the chain rule, rank $(f \circ \varphi)' = r$, so

$$\frac{\partial h_j}{\partial w_k} = 0, \quad \text{for all} \quad j, k \geq r + 1.$$

Thus, the h_j's do not depend on the variables $w_{r+1}, \cdots, w_n$ and their restrictions to the first r components define a mapping $h : D^r \to \mathbb{C}^{m-r}$. The mapping

$$\gamma : D^r \times \mathbb{C}^{m-r} \to D^r \times \mathbb{C}^{m-r}$$

$$(u, v) \mapsto (u, v - h(u))$$

is bijective and γ' has the following matrix:

$$\begin{pmatrix} I_r & 0 \\ * & I_{m-r} \end{pmatrix}.$$

From the inverse mapping theorem, it follows that γ is biholomorphic. Now choose a polydisc D^{m-r} in $\mathbb{C}^{m-r}$ so large that

$$(\gamma \circ f \circ \varphi)(D^n) \subset D^r \times D^{m-r} =: D^m;$$

for $V := \gamma^{-1}(D^m)$ and $\psi := \gamma|_V$. It follows that

$$(\psi \circ f \circ \varphi)(w) = \gamma(w_1, \cdots, w_r, h_{r+1}(w), \cdots, h_m(w)) =$$
$$= (w_1, \cdots, w_r, 0, \cdots, 0) \qquad\qquad = \chi(w).$$

□

The rank theorem has a real version for smooth mappings (see [16]), which we shall call the real rank theorem and we shall refer to the holomorphic version which we have just proved as the complex rank theorem.

An important class of holomorphic mappings are proper holomorphic mappings. A function $f : \Omega_1 \to \Omega_2$, from a topological space $\Omega \subset \mathbb{C}^n$ to a topological space $\Omega \subset \mathbb{C}^m$, is said to be *proper* if, for each compact $K \subset \Omega_2$, the inverse image $f^{-1}(K)$ is also compact. Let us say that a sequence z_j, $j = 1, 2, \cdots$, in a metric space (M, d) tends to a set $A \subset M$, if $d(z_j, A) \to 0$, in which case, we write $z_j \to A$.

Problem 28. Show that if $\Omega_1 \subset \mathbb{C}^n$ and $\Omega_2 \subset \mathbb{C}^m$ are bounded, then a continuous mapping $f : \Omega_1 \to \Omega_2$ is proper if and only if, for each sequence z_j, $j = 1, 2, \ldots$, in Ω_1 which tends to the boundary $\partial\Omega_1$, the sequence $f(z_j)$, $j = 1, 2, \cdots$ tends to the boundary $\partial\Omega_2$.

Clearly *every homeomorphism is proper and consequently every biholomorphic mapping is proper*. The unit disc $\mathbb{D}$ and the complex plane $\mathbb{C}$ are not biholomorphic. In fact, Liouville's theorem states that there is no (nonconstant) holomorphic function $f : \mathbb{C} \to \mathbb{D}$. However, there are obviously many nonconstant holomorphic functions in the other direction $f : \mathbb{D} \to \mathbb{C}$.

Problem 29. Show that there is no proper holomorphic function $f : \mathbb{D} \to \mathbb{C}$.

Chapter 8
Plurisubharmonic Functions

Abstract Potential theory enters the scene via plurisubharmonic functions. They form a bridge between potential theory and complex analysis, for they include the important functions $\Re f$ and $\log|f|$, when f is holomorphic. They provide a powerful tool because of their great flexibility. They are not as rigid as holomorphic functions. For example, a plurisubharmonic can be zero on an open subset of a domain, without being zero everywhere on the domain.

Problem 30. If f is a holomorphic function of several complex variables, then the real part of f is harmonic.

Recall that in one complex variable, there is a sort of converse. If $u(x, y)$ is a harmonic function of two real variables, then u is locally the real part of a holomorphic function $f(z) = u(x, y) + iv(x, y)$.

In several variables, there is no such converse. Consider the function $u(x_1, y_1, x_2, y_2) = x_1^2 - x_2^2$, where $z_1 = x_1 + iy_1$ and $z_2 = x_2 + iy_2$. Then, u is harmonic, but suppose there were locally a holomorphic function $f(z_1, z_2)$ such that $f = u + iv$. Then, for fixed z_2, the function $f_1(z_1) = f(z_1, z_2)$ would be holomorphic and hence the real part $u_1(x_1, y_1) = x_1^2 - x_2^2$ would be harmonic in (x_1, y_1) which it is not.

A *complex line* in $\mathbb{C}^n$ is a set of the form $\ell = \{z : z = a + \lambda b, \lambda \in \mathbb{C}\}$, where a and b are fixed points in $\mathbb{C}^n$, with $b \neq 0$. Let us say that ℓ is the complex line through a in the "direction" b. Let $e^1, \cdots, e^n$ be the standard basis of $\mathbb{C}^n$. Thus, the coordinates of e^j are given by the Kronecker delta δ_k^j. The complex line through a in the direction of e^j is called the complex line through a in the direction of the j-th coordinate.

If Ω is an open set in $\mathbb{C}^n$, we defined f to be holomorphic in Ω if $f \in C^1(\Omega)$ and f is holomorphic in each variable, that is, if the restriction of f to $\ell \cap \Omega$ is holomorphic for each complex line ℓ in the direction of a coordinate. In fact, the following theorem asserts that the restriction to *every* complex line is holomorphic.

Theorem 34. *Let Ω be open in $\mathbb{C}^n$ and $f \in C^1(\Omega)$. Then, $f \in \mathcal{O}(\Omega)$ if and only if the restriction of f to $\ell \cap \Omega$ is holomorphic, for each complex line ℓ.*

© Springer International Publishing Switzerland 2014
P.M. Gauthier, *Lectures on Several Complex Variables*,
DOI 10.1007/978-3-319-11511-5_8

Proof. The restriction of f to a complex line $\{z = a + \lambda b : \lambda \in \mathbb{C}\}$ is the function $f(a + \lambda b)$. If the restrictions are holomorphic, then f is by definition holomorphic. Conversely, it follows from Problem 24 that if $f \in \mathcal{O}(\Omega)$, then the restriction of f to $\ell \cap \Omega$ is holomorphic, for all complex lines ℓ. $\square$

As mentioned in the introduction, we may drop the assumption that f is smooth. This very powerful fact is known as Hartogs' theorem on separate holomorphy.

Theorem 35. *Let Ω be open in $\mathbb{C}^n$ and f a function on Ω. Then, $f \in \mathcal{O}(\Omega)$ if and only if the restriction of f to $\ell \cap \Omega$ is holomorphic, for all complex lines ℓ in coordinate directions.*

A real-valued function u defined in an open subset Ω of $\mathbb{C}^n$ is said to be *pluriharmonic* in Ω if $u \in C^2(\Omega)$ and the restriction of u to $\ell \cap \Omega$ is harmonic for each complex line ℓ. Unlike the holomorphic situation, this is *not* equivalent to being harmonic in each *coordinate* direction.

Let Ω be an open set in $\mathbb{C}^n$. For $u \in C^2(\Omega)$, the Hermitian matrix

$$L_u = \left(\frac{\partial^2 u}{\partial z_j \, \partial \bar{z}_k} \right)$$

is called the *complex Hessian matrix* of u. We use the letter L for the complex Hessian, because the letter H is already being used for the real Hessian and because the quadratic form associated to the complex Hessian is usually called the *Levi form*. Since $L_u(z)$ is Hermitian, the eigenvalues are all real.

We shall see later that a real function $u \in C^2(\Omega)$ is pluriharmonic in Ω if and only if its complex Hessian matrix vanishes identically, $L_u = 0$, that is, if and only if u satisfies the system of differential equations

$$\frac{\partial^2 u}{\partial z_j \, \partial \bar{z}_k}(z) = 0, \quad \forall z \in \Omega.$$

In real form, this system of equations becomes

$$\frac{\partial^2 u}{\partial x_j \, \partial x_k} + \frac{\partial^2 u}{\partial y_j \, \partial y_k} = 0, \qquad \frac{\partial^2 u}{\partial x_j \, \partial y_k} - \frac{\partial^2 u}{\partial x_k \, \partial y_j} = 0. \tag{8.1}$$

We may now characterize real parts of holomorphic functions.

Theorem 36. *The real part of any holomorphic function is pluriharmonic. Conversely, every pluriharmonic function is locally the real part of a holomorphic function.*

Proof. It is an immediate consequence of Theorem 34 that the real part of a holomorphic function is pluriharmonic.

To show the converse, it is sufficient to show that any function u pluriharmonic in a polydisc $\mathcal{D}$ is the real part of a holomorphic function therein.

We shall use the Poincaré lemma which asserts that in a convex domain, every closed form is exact (see, for example, [16, Theorem 10.39]).

We wish to show that there exists a function v such that $f = u + iv$ is holomorphic. If u did have such a conjugate function v, we could write

$$v(z) - v(a) = \int_a^z dv.$$

Since conjugate functions are only determined up to additive imaginary constants, we could even assume that $v(a) = 0$. From the Cauchy–Riemann equations, we would have

$$dv = \sum_k \left(\frac{\partial v}{\partial x_k} dx_k + \frac{\partial v}{\partial y_k} dy_k \right) = \sum_k \left(-\frac{\partial u}{\partial y_k} dx_k + \frac{\partial u}{\partial x_k} dy_k \right) = *du.$$

Now dv is undefined, since we are trying to prove the existence of v, but the conjugate differential $*du$ of u is well defined by the last equality. Set $\omega = *du$. If we can show that ω is an exact differential, that is, that there is in fact a C^1-function v such that $dv = \omega$, then u and v will satisfy the Cauchy–Riemann equations and so $f = u + iv$ will indeed be holomorphic.

Since we are working in a polydisc, which is thus a convex domain, we need only check that the differential form ω is closed. By the Poincaré lemma it will then be exact.

$$d\omega = -\sum_{j,k} \frac{\partial^2 u}{\partial x_j \partial x_k} dx_j \wedge dx_k + \sum_{j,k} \frac{\partial^2 u}{\partial y_j \partial x_k} dy_j \wedge dy_k +$$

$$+ \sum_{j,k} \left(\frac{\partial^2 u}{\partial x_j \partial x_k} + \frac{\partial^2 u}{\partial y_j \partial y_k} \right) dx_j \wedge dy_k.$$

The first sum is zero because

$$\frac{\partial^2 u}{\partial x_j \partial x_k} = \frac{\partial^2 u}{\partial x_k \partial x_j} \qquad \text{while} \qquad dx_j \wedge dx_k = -dx_k \wedge dx_j.$$

The second sum is zero for a similar reason. The third sum is zero because by (8.1) the terms are zero. Thus $d\omega = 0$ and the proof is complete. □

Pluriharmonic functions have the following local representation.

Problem 31. A function u defined in an open subset Ω of $\mathbb{C}^n$ is pluriharmonic in Ω if and only if it can be locally represented in the form

$$u = f + \overline{g},$$

where f and g are holomorphic.

Recall the definition of a subharmonic function. A function $u : G \to [-\infty, +\infty)$, defined on an open subset $G \subset \mathbb{R}^n$, is said to be *subharmonic* if it satisfies the following: u is upper semicontinuous, u is not identically $-\infty$ on any component of G, and, for each closed ball $\overline{B} \subset G$ and real-valued continuous function h on $\overline{B}$ which is harmonic on B, if $u \leq h$ on ∂B, then $u \leq h$ in B. The following theorem tells us that, for continuous functions u, we don't need to check all functions h. It is sufficient to check the solution P_B^u of the Dirichlet problem on B, with boundary values given by u.

Theorem 37. *A continuous function $u : G \to \mathbb{R}$ is subharmonic if and only $u|_B \leq P_B^u$, for each closed ball $\overline{B} \subset G$.*

Proof. Suppose u is subharmonic and $\overline{B} \subset \Omega$. Since P_B^u is continuous on $\overline{B}$ and harmonic on B and since $P_B^u = u$ on ∂B, it follows from the definition of subharmonicity that $u|_B \leq P_B^u$.

Conversely, suppose $u|_B \leq P_B^u$ and let h be a continuous function on $\overline{B}$, which is harmonic in B and such that $u \leq h$ on ∂B. Since $P_B^u = u$ on ∂B, we have $P_B^u \leq h$ on ∂B. By the maximum principle for harmonic functions, $P_B^u \leq h$ on B. Thus, on B, it follows that $u|_B \leq P_B^u \leq h$, so $u|_B \leq h$. Hence, u is subharmonic on Ω. $\square$

For basic properties of subharmonic functions, we refer to [3]. We recall one property which is very important. Namely, an upper-semicontinuous function u on a domain $\Omega \subset \mathbb{R}^n$ is subharmonic if and only if it satisfies the *mean value inequality* also called the *sub-mean-value property*, which states that for each closed ball $\overline{B} \subset \Omega$, the value of u at the center of B is dominated by its mean value on the boundary of B.

Remark. If $\mathcal{F}$ is a locally upper bounded family of subharmonic functions on an open set $\Omega \subset \mathbb{R}^n$ and if $u = \sup_{v \in \mathcal{F}} v$ is upper semicontinuous, then u is subharmonic. Indeed, u is upper semicontinuous and if u were identically $-\infty$ on some component of Ω, the same would be true of every $v \in \mathcal{F}$, contradicting the subharmonicity of v. Now let $\overline{B}$ be a closed ball in Ω and h a real-valued continuous function on $\overline{B}$ which is harmonic on B. If $u \leq h$ on ∂B, then the same is true of each $v \in \mathcal{F}$. Thus, $v \leq h$ in B. Therefore $u = \sup_{v \in \mathcal{F}} v \leq h$ in B. We have shown that u is subharmonic in Ω.

Example. If Ω is an open set in $\mathbb{R}^2 = \mathbb{C}$, and $\delta_\Omega(z)$ is the distance of z from the boundary $\partial \Omega$, then $-\log \delta_\Omega$ is subharmonic on Ω. Indeed, this follows from the preceding remark. First of all, let us show that $-\log \delta_\Omega$ is continuous on Ω. It is sufficient to show that δ_Ω is continuous. Fix $z \in \Omega$ and $\epsilon > 0$. Suppose $\zeta \in \Omega$ and $|z - \zeta| < \epsilon$. We have

$$\delta_\Omega(\zeta) = \inf_{a \in \partial\Omega} |\zeta - a| \leq \inf_{a \in \partial\Omega} (|z - a| + |\zeta - a| - |z - a|) \leq$$

$$\inf_{a \in \partial\Omega} (|z - a| + ||\zeta - a| - |z - a||) \leq \inf_{a \in \partial\Omega} (|z - a| + |\zeta - z|) \leq$$

$$\inf_{a \in \partial\Omega} (|z - a| + \epsilon) \leq \inf_{a \in \partial\Omega} |z - a| + \epsilon = \delta_\Omega(z) + \epsilon.$$

The same argument shows that $\delta_\Omega(z) \leq \delta_\Omega(\zeta) + \epsilon$. We have shown that $|z - \zeta| < \epsilon$ implies that $|\delta_\Omega(z) - \delta_\Omega(\zeta)| \leq \epsilon$, so δ_Ω is continuous on Ω. Since

$$-\log \delta_\Omega(z) = \sup_{a \in \partial\Omega} (-\log|z - a|), \quad z \in \Omega,$$

it follows from the preceding remark that $-\log \delta_\Omega$ is subharmonic.

A function u defined in an open subset Ω of $\mathbb{C}^n$, and taking values in $[-\infty, +\infty)$, is said to be *plurisubharmonic* in Ω if u is upper semicontinuous, u is not identically $-\infty$ on any component of Ω, and, for each complex line ℓ, the restriction of u to each component of $\ell \cap \Omega$ is subharmonic or identically $-\infty$.

Theorem 38. *A continuous function $u : \Omega \to \mathbb{R}$ is plurisubharmonic if and only if $u|_D \leq P_D^u$, for each closed complex disc $\overline{D} \subset \Omega$.*

Proof. By Theorem 37, the restriction of u to a component G of $\ell \cap \Omega$ is subharmonic if and only if $u|_D \leq P_D^u$, for each closed disc $\overline{D} \subset \ell \cap \Omega$. Now the family of closed complex discs $\overline{D} \subset \Omega$ is the same as the family of closed discs $\overline{D} \subset \ell \cap \Omega$, over all complex lines ℓ. $\qquad\square$

Similarly, one can define plurisuperharmonic functions, and it is easy to see that a function u is plurisuperharmonic if and only if $-u$ is plurisubharmonic.

Problem 32. If f is holomorphic, then $|f|^p$, for $p > 0$, and $\log|f|$ are plurisubharmonic.

Example. In $\Omega = \mathbb{C}^2 \setminus \{0\}$ the function $u(z) = -\log|z|$ is not plurisubharmonic. To see this, we show that the restriction of u to the complex line $\ell = \{(1, z_2) : z_2 \in \mathbb{C}\}$ is not subharmonic, because it does not satisfy the mean value inequality at the point $a = (1, 0)$. Consider the disc $D = \{z = (1, z_2) \in \ell : |z_2| = 1\}$. For $z \in \partial D$, $|z|^2 = 1^2 + |z_2|^2 = 2$. Thus, $|z| > |a|$, since $|a| = 1$ and $u(a) = -\log|a| > -\log|z| = u(z)$. Therefore, $u(a)$ is greater than its average on the boundary of the disc D and so does not satisfy the mean value inequality at a. Thus, $u|\ell$ is not subharmonic and consequently u is not plurisubharmonic in Ω.

Problem 33. For an open set $\Omega \subset \mathbb{C}^n$, the class of functions $\log|f|$, $f \in \mathcal{O}(\Omega)$, contains the class of pluriharmonic functions on Ω.

Problem 34. Upper-semicontinuous functions, and in particular subharmonic and plurisubharmonic functions, are locally upper bounded.

Many properties of plurisubharmonic functions follow from the following theorem.

Theorem 39. *Plurisubharmonic functions are subharmonic.*

Proof. Let v be plurisubharmonic in a domain Ω. To show that v is subharmonic, it is sufficient to show that v locally satisfies the sub-mean-value property. Fix $a \in \Omega$. If $v(a) = -\infty$, then the sub-mean-value property is obvious at the point a. Suppose $v(a)$ is finite and let B be a ball centered at a, whose closure is contained in Ω.

We may assume that $a = 0$ and B is the unit ball. Since v is upper semicontinuous it is locally upper bounded and so the integral of v on the unit sphere S makes sense (with possibly the value $-\infty$). Integration by slices (see [17]) yields

$$\int_S v d\sigma = \int_s d\sigma(\zeta) \frac{1}{2\pi} \int_{-\pi}^{\pi} v(e^{i\theta}\zeta) d\theta \geq v(0),$$

where σ is normalized Lebesgue measure on the sphere S and the last inequality is just the sub-mean-value property for the subharmonic function $v(\lambda\zeta)$, as a function of λ, for $\zeta \in S$ fixed. We have shown the sub-mean-value property for upper-semicontinuous function v and therefore v is subharmonic. $\square$

We recall another notion from the theory of subharmonic functions. A subset E of a domain $\Omega \subset \mathbb{R}^n$ is said to be a *polar* set, if $E \subset u^{-1}(-\infty)$, for some function u subharmonic on Ω.

Theorem 40. *If u is plurisubharmonic and $\lambda > 0$, then λu is plurisubharmonic, and if u_1 and u_2 are plurisubharmonic, then so are $u_1 + u_2$ and $\max\{u_1, u_2\}$.*

Proof. Suppose u_1 and u_2 are plurisubharmonic on a domain Ω, ℓ is a complex line and U is a component of $\ell \cap \Omega$. If u_1 and u_2 are subharmonic on U, then $u_1 + u_2$ is also subharmonic on U. If u_1 or u_2 is identically $-\infty$ on U, then the same is true for $u_1 + u_2$. To see that $u_1 + u_2$ is plurisubharmonic on Ω, there only remains to show that $u_1 + u_2 \not\equiv -\infty$. Now u_1 and u_2 are both subharmonic on Ω. The sets $u_1^{-1}(-\infty)$ and $u_2^{-1}(-\infty)$ are polar sets. The union of two polar sets is again a polar set, so $u_1 + u_2 \not\equiv -\infty$. Thus, $u_1 + u_2$ is plurisubharmonic.

We leave the proof that λu and $\max\{u_1, u_2\}$ are plurisubharmonic to the reader. $\square$

We would like to pass from the maximum of two plurisubharmonic functions to the supremum of an upper bounded family of plurisubharmonic functions. We cannot quite do this—but almost. For a function u defined on a domain Ω and taking its values in $[-\infty, +\infty)$, we define the (upper) regularization u^* as

$$u^*(z) = \lim_{\epsilon \to 0} \sup_{|\zeta-z|<\epsilon} u(\zeta).$$

Theorem 41. *Let $\mathcal{U}$ be a locally upper bounded family of plurisubharmonic functions on a domain Ω and set $u = \sup\{v : v \in \mathcal{U}\}$. Then, u^* is plurisubharmonic on Ω.*

Proof. Certainly, u^* takes its values in $[\infty, +\infty)$, since $\mathcal{U}$ is locally upper bounded. Also, u^* cannot be identically $-\infty$ on Ω, for then, so would every $v \in \mathcal{U}$. Let ℓ be a complex line and $\overline{D}$ be a closed disc in $\ell \cap \Omega$. To show that $u^*|_D$ is subharmonic, it is sufficient to show that, if h is harmonic on $\overline{D}$ and $u^* \leq h$ on ∂D, then $u^* \leq h$ in D. For every $v \in \mathcal{U}$, we have $v \leq u^* \leq h$ on ∂D. Since $v|_D$ is subharmonic, $v \leq h$ on D. Hence, $u \leq h$ on D and so $u^* \leq h^* = h$ on D. Thus, $u^*|_D$ is subharmonic. $\square$

Plurisubharmonic functions bear a resemblance to convex functions. We present an example of this resemblance by first characterizing convex C^2-functions and then giving a similar characterization for plurisubharmonic C^2-functions.

Our discussion of convex functions follows the presentation in Fleming [5].

Recall that a real-valued function u defined on a convex open set Ω in $\mathbb{R}^n$ is said to be *convex* if for each $a, b \in \Omega$, and each $t \in [0, 1]$,

$$u(ta + (1 - t)b) \le tu(a) + (1 - t)u(b). \tag{8.2}$$

The function is called *strictly* convex if

$$u(ta + (1 - t)b) < tu(a) + (1 - t)u(b),$$

for each $a \ne b$ in Ω and each $t \in (0, 1)$. Let us say that the function is *midpoint convex*, if

$$u\left(\frac{a + b}{2}\right) \le \frac{u(a) + u(b)}{2},$$

for each $a, b \in \Omega$.

Theorem 42. *Let u be a real-valued function defined on a convex subset Ω of $\mathbb{R}^n$. If u is continuous, then u is convex if and only if it is midpoint convex.*

Proof. Due to the symmetry between a and b and between t and $1 - t$, we only need to prove (8.2) for $t \in (0, 1/2)$. Also, (8.2) is equivalent to

$$u(x + t(y - x)) \le u(x) + t(u(y) - u(x)), \tag{8.3}$$

We first show (8.3) for all t of the form

$$t = j2^{-n}, j = 0, \cdots, 2^n. \tag{8.4}$$

We shall proceed by induction on n. For $n = 0$, the assertion is trivial and for $n = 1$ it follows immediately from the definition of convexity. Suppose now that we have shown (8.3) for n. Let $t = j2^{-(n+1)}$. Now, setting $w = 2^{-1}(x + y)$, we have

$$u(x + j2^{-(n+1)}(y - x)) = u(x + j2^{-n}\frac{y - x}{2}) \le u(x + j2^{-n}(w - x)) \le$$

$$u(x) + j2^{-n}(u(w) - u(x)) = u(x) + j2^{-n}(u(\frac{x + y}{2}) - u(x)) \le$$

$$\le u(x) + j2^{-(n+1)}(u(y) - u(x)),$$

by the inductive hypothesis, which is valid provided $j2^{-n} \le 1$. For $j = 2^n + 1, \cdots, 2^{n+1}$, we set $k = 2^{n+1} - j$. Then,

$$x + j2^{n+1}(y - x) = y + k2^{n+1}(x - y)$$

and we are back in the justifiable case. By induction, we have shown (8.3) for every t of the form (8.4). Now, we fix $x, y \in \Omega$ and set

$$\varphi(t) = u(x + t(y - x)) - u(x) - t(u(y) - u(x)),$$

for $t \in (0, 1)$. By hypothesis, φ is continuous and we have proven that $\varphi(t) \leq 0$, for the dense set of t of the form (8.4). Therefore $\varphi(t) \leq 0$, for all $t \in (0, 1)$ which establishes (8.3) and ends the proof. $\square$

Corollary 43. *Let u be a real-valued function defined on a convex subset Ω of $\mathbb{R}^n$. If u is continuous, then u is convex if and only if*

$$u(p_1 x_1 + \cdots + p_m x_m) \leq p_1 u(x_1) + \cdots + p_m u(x_m), \qquad (8.5)$$

whenever $x_1, \cdots, x_m \in \Omega$ and $0 \leq p_j \leq 1$, with $p_1 + \cdots + p_m = 1$.

Proof. The proof is by induction. For $m = 1$ the assertion is trivial and for $m = 2$ it is the definition of convexity. Suppose the assertion is true for m and let x_j and p_j be as in the theorem with $j = 1, \cdots, m + 1$. We may assume that $0 < p_{m+1} < 1$. We note that

$$p_1 x_1 + \cdots + p_m x_m = (1 - p_{m+1})y,$$

where

$$y = \frac{p_1 x_1 + \cdots p_m x_m}{1 - p_{m+1}} = \sum_{j=1}^{m} \frac{p_j}{1 - p_{m+1}} x_j,$$

and

$$\sum_{j=1}^{m} \frac{p_j}{1 - p_{m+1}} = 1.$$

Therefore, since Ω is convex, $y \in \Omega$. By the theorem,

$$u(p_1 x_1 + \cdots + p_m x_m) =$$

$$u\left((1 - p_{m+1})y + p_{m+1}x_{m+1}\right) \leq (1 - p_{m+1})u(y) + p_{m+1}u(x_{m+1}) =$$

$$(1 - p_{m+1})u\left(\sum_{j=1}^{m} \frac{p_j}{1 - p_{m+1}} x_j\right) + p_{m+1}u(x_{m+1}) \leq$$

$$(1 - p_{m+1})\sum_{j=1}^{m} \frac{p_j}{1 - p_{m+1}} u(x_j) + p_{m+1}u(x_{m+1}) = p_1 u(x_1) + \cdots + p_{m+1}u(x_{m+1}),$$

where the next-to-last equality is by the induction hypothesis. $\square$

Having characterized continuous convex functions, we now characterize differentiable convex functions.

Theorem 44. *Let u be a real-valued function defined on a convex open subset Ω of $\mathbb{R}^n$. If u is differentiable, then u is convex if and only if*

$$u(y) \geq u(x) + \nabla u(x) \cdot (y - x), \tag{8.6}$$

for every $x, y \in \Omega$.

Proof. The condition in the theorem certainly corresponds to the intuitive notion of a function being convex if its graph $\{(y, u(y)) : y \in \Omega\}$ in $\mathbb{R}^{n+1}$ is concave, for the condition says that the graph lies above the tangent space to the graph at $(x, u(x))$, for each $x \in \Omega$.

Suppose u is convex in Ω and let $x, y \in \Omega$. Let $h = y - x$ and $t \in (0, 1)$. From the convexity of u,

$$u(x + th) \leq tu(x + h) + (1 - t)u(x).$$

This can be rewritten as

$$u(x + th) - u(x) \leq t[u(x + h) - u(x)].$$

Subtracting $t\nabla u(x) \cdot h$ from both sides and dividing by t, we have

$$\frac{u(x + th) - u(x) - t\nabla u(x) \cdot h}{t} \leq u(x + h) - u(x) - \nabla u(x) \cdot h.$$

Since u is differentiable, the left side approaches 0 when $t \to 0^+$. Therefore we have (8.6).

Conversely, assume that (8.6) holds for every $x, y \in \Omega$. Let $x_1, x_2 \in \Omega$, $x_1 \neq x_2$. Set

$$x = \frac{x_1 + x_2}{2}, \quad h = x_1 - x.$$

Then $x_2 = x - h$. From (8.6) we have

$$u(x_1) \geq u(x) + \nabla u(x) \cdot h,$$
$$u(x_2) \geq u(x) + \nabla u(x) \cdot (-h).$$

Adding the inequalities, we arrive at

$$u(x_1) + u(x_2) \geq 2u(x) \quad \text{or} \quad \frac{u(x_1) + u(x_2)}{2} \geq u\left(\frac{x_1 + x_2}{2}\right).$$

Hence, u is convex. $\qquad\qquad\square$

For a real-valued C^2-function u, let H_u denote the Hessian matrix of u. We write $H_u \geq 0$ to mean that the associated quadratic form is positive semi-definite. Having characterized differentiable convex functions, we now characterize C^2-convex functions.

Theorem 45. *Let u be a real-valued function defined on a convex open subset Ω of $\mathbb{R}^n$. If $u \in C^2(\Omega)$, then u is convex if and only if $H_u \geq 0$, that is,*

$$\left(\frac{\partial^2 u}{\partial x_j \, \partial x_k} \right) \geq 0. \tag{8.7}$$

Proof. We need to show that u is convex if and only if

$$\sum_{j,k=1}^{n} \frac{\partial^2 u}{\partial x_j \, \partial x_k}(x) h_j h_k \geq 0, \quad \text{for all } h \in \mathbb{R}^n, x \in \Omega.$$

Since Ω is convex, Taylor's formula is valid for every pair of points $x, y \in \Omega$:

$$u(y) = u(x) + \nabla u(x) \cdot h + \sum_{|\alpha|=2} \frac{1}{\alpha!} \frac{\partial^2 u}{\partial x^\alpha}(x + sh) h^\alpha = \tag{8.8}$$

$$u(x) + \nabla u(x) \cdot h + \sum_{j=k} \frac{1}{2} \frac{\partial^2 u}{\partial x_j^2}(x + sh) h_j^2 + \sum_{j<k} \frac{\partial^2 u}{\partial x_j \partial x_k}(x + sh) h_j h_k =$$

$$u(x) + \nabla u(x) \cdot h + \frac{1}{2} \sum_{j,k=1}^{n} \frac{\partial^2 u}{\partial x_j \partial x_k}(x + sh) h_j h_k,$$

where $s \in (0, 1)$ and $h = y - x$. So, if $Q_u(x, h)$ is the quadratic form associated to the Hessian $H_u(x)$, we have

$$u(y) = u(x) + \nabla u(x) \cdot h + \frac{1}{2} Q_u(x + sh, h). \tag{8.9}$$

Now, to prove the theorem, suppose (8.7) holds for each $z \in \Omega$. Then, for $z = x + sh$, it follows from (8.9) that

$$u(y) \geq u(x) + \nabla u(x) \cdot h.$$

Hence, u satisfies (8.6) and consequently by Theorem 44 u is convex.

In the other direction, if (8.7) fails at some point $x \in \Omega$, then $Q_u(x_0, h_0) < 0$ for some $x_0 \in \Omega$ and some $h_0 \neq 0$. Since $u \in C^2(\Omega)$, the function $Q_u(\cdot, h_0)$ is continuous in Ω. Thus, there is a $\delta > 0$ such that $Q_u(y, h_0) < 0$ for every y in the δ-neighborhood of x_0. Let $h = ch_0$, with c so small that $|h| < \delta$, and set $x = x_0 + h$. Fix an $s \in (0, 1)$. Since $Q_u(x_0 + sh, \cdot)$ is quadratic,

$$Q_u(x_0 + sh, h) = c^2 Q_u(x_0 + sh, h_0) < 0.$$

From (8.8)

$$u(x) < u(x_0) + \nabla u(x_0) \cdot h.$$

By Theorem 44, u is not convex in Ω.   □

We now state an analogous characterization of plurisubharmonic functions.

Theorem 46. *Let Ω be an open set in $\mathbb{C}^n$. A real-valued function $u \in C^2(\Omega)$ is plurisubharmonic in Ω if and only if $L_u \geq 0$, that is,*

$$\left(\frac{\partial^2 u}{\partial z_j \, \partial \overline{z}_k} \right) \geq 0.$$

Proof. If ℓ is a complex line meeting Ω in $\mathbb{C}^n$, we denote by u_ℓ the restriction of u to $\ell \cap \Omega$. Then, $u_\ell(\lambda)$ is a C^2-function on $\ell \cap \Omega$.

Suppose $L_u \geq 0$, and ℓ is a complex line meeting Ω. Then,

$$\frac{\partial^2 u_\ell}{\partial \lambda \, \partial \overline{\lambda}} = \sum_j \frac{\partial}{\partial \lambda} \left(\frac{\partial u}{\partial \overline{z}_j} \frac{\overline{\partial \ell_j}}{\partial \lambda} \right) = \sum_j \left(\left(\frac{\partial}{\partial \lambda} \frac{\partial u}{\partial \overline{z}_j} \right) \frac{\overline{\partial \ell_j}}{\partial \lambda} + \frac{\partial u}{\partial \overline{z}_j} \left(\frac{\partial}{\partial \lambda} \frac{\overline{\partial \ell_j}}{\partial \lambda} \right) \right) =$$

$$\sum_{j,k} \frac{\partial^2 u}{\partial z_k \overline{z}_j} \frac{\partial \ell_k}{\partial \lambda} \frac{\overline{\partial \ell_j}}{\partial \lambda} \geq 0.$$

Thus, u_ℓ is subharmonic.

Conversely, if u is plurisubharmonic and $w \neq 0$, set $\ell = a + \lambda w$. Then $\partial \ell_j / \partial \lambda = w_j$. Thus,

$$\sum \frac{\partial^2 u}{\partial z_j \, \partial \overline{z}_k} w_j \overline{w}_k = \frac{\partial^2 u_\ell}{\partial \lambda \, \partial \overline{\lambda}} \geq 0,$$

so $L_u \geq 0$.   □

A real-valued function $u \in C^2(\Omega)$ is said to be *strictly* plurisubharmonic in Ω if its Levi form is strictly positive, that is, $L_u > 0$ on Ω.

Corollary 47. *Let Ω be an open set in $\mathbb{C}^n$. A real-valued function $u \in C^2(\Omega)$ is plurisubharmonic in Ω if and only if, at each point $z \in \Omega$, the eigenvalues of $L_u(z)$ are nonnegative. It is strictly plurisubharmonic if and only if, at each point $z \in \Omega$, the eigenvalues of $L_u(z)$ are positive.*

It is useful to recall here that the eigenvalues λ of a real matrix A are the roots of the characteristic polynomial $\det(A - \lambda I)$.

Corollary 48. *A real function in $C^2(\Omega)$ is pluriharmonic if and only if $L_u = 0$ on Ω and, equivalently, if and only if it is both plurisubharmonic and plurisuperharmonic.*

In one complex variable, a useful notion of small sets is that of polar sets. A subset E of a domain $G \subset \mathbb{C}$ is *polar* if there is a subharmonic function u ($\not\equiv -\infty$) on G such that $u(z) = -\infty$ on E. Similarly, a subset E of a domain $G \subset \mathbb{C}^n$ is *pluripolar* if there is a plurisubharmonic function u ($\not\equiv -\infty$) on G such that $u(z) = -\infty$ on E. As in one variable, pluripolarity is a useful notion of smallness for sets in $\mathbb{C}^n$.

Chapter 9
The Dirichlet Problem

Abstract The natural Dirichlet problem in $\mathbf{C}^n$ is not the classical one for solutions of the Laplace equation but rather for solutions of the Monge–Ampère equation. We present the solution of this Dirichlet problem in a ball.

The classical Dirichlet problem is the following. Given a bounded open subset Ω of $\mathbb{R}^n$ and a continuous function φ on the boundary $\partial\Omega$, find a harmonic function u in Ω having boundary values φ. That is, find a function u continuous on $\overline{\Omega}$ such that $\Delta u = 0$ in Ω and $u = \varphi$ on $\partial\Omega$. One way of attacking the Dirichlet problem is via the method of Perron using subharmonic functions.

Harmonic functions in $\mathbb{C}^n$ have the serious drawback that harmonicity is not preserved by biholomorphic change of coordinates. That is, *if u is harmonic and L is a linear change of coordinates in $\mathbb{C}^n$, then $u \circ L$ need not be harmonic.* For the purposes of complex analysis in several variables, it would seem more appropriate to find a solution to the Dirichlet problem which is pluriharmonic. The class of pluriharmonic functions is a more restricted class than the class of harmonic functions. For the Dirichlet problem, the class of pluriharmonic functions is in fact too restricted. There exist continuous functions φ on the boundary of such smooth domains as the ball, for which there is no solution to the Dirichlet problem in the class of pluriharmonic functions. We seek to enlarge the class of pluriharmonic functions sufficiently to solve the Dirichlet problem while retaining the property that this larger class will be preserved by complex change of coordinates. A solution is provided in terms of the complex Monge–Ampère equation.

The complex Monge–Ampère equation is the nonlinear partial differential equation

$$\det H(u) = 0.$$

Since $H(u) = 0$ for pluriharmonic functions, it is trivial that the class of solutions to the complex Monge–Ampère equation contains the pluriharmonic functions.

Problem 35. The class of solutions to the complex Monge–Ampère equation is preserved by linear change of coordinates.

© Springer International Publishing Switzerland 2014
P.M. Gauthier, *Lectures on Several Complex Variables,*
DOI 10.1007/978-3-319-11511-5_9

However, since the Monge–Ampère operator $u \mapsto \det H(u)$ is a nonlinear operator, it is not surprising that the sum of two solutions need not be a solution. For example, in $\mathbb{C}^2$, consider the functions $u_1(z) = |z_1|^2 = z_1 \bar{z}_1$ and $u_2(z) = |z_2|^2 = z_2 \bar{z}_2$. Then,

$$H(u_1) = \begin{pmatrix} 1 & 0 \\ 0 & 0 \end{pmatrix} \quad \text{and} \quad H(u_2) = \begin{pmatrix} 0 & 0 \\ 0 & 1 \end{pmatrix},$$

so u_1 and u_2 are solutions to the Monge–Ampère equation, but $u_1 + u_2$ is not.

Just as the Perron method uses subharmonic functions to find a harmonic solution h to the Dirichlet problem, it is possible in the ball to use the Perron method with plurisubharmonic functions to find a solution u to the Dirichlet problem which is plurisubharmonic and satisfies the complex Monge–Ampère equation.

Theorem 49. *The Dirichlet problem for the complex Monge–Ampère equation has a solution in the ball.*

Proof. Let $\mathbb{B}$ be a ball in $\mathbb{C}^n$ and $\varphi \in C(\partial\mathbb{B})$. Denote by $\mathcal{U}$ the family of all plurisubharmonic functions v in $\mathbb{B}$ which are dominated by φ at the boundary $\partial\mathbb{B}$. That is,

$$\limsup v(z)_{z \to \zeta} \leq \varphi(\zeta) \quad \forall \zeta \in \partial\mathbb{B}.$$

Now set

$$\omega(z) = \sup_{v \in \mathcal{U}} v(z).$$

By Theorem 41 the regularization ω^* is plurisubharmonic in $\mathbb{B}$. One can show that it is continuous in $\overline{\mathbb{B}}$ and satisfies

$$\det H(\omega^*) = 0 \quad \text{and} \quad \omega^* |_{\partial\mathbb{B}} = \varphi.$$

Of course, if $\omega*$ is not differentiable, one must give meaning to the Monge–Ampère equation $\det H(\omega^*) = 0$. This can be done using the theory of currents, but this is beyond the scope of these lectures. $\square$

This Monge–Ampère solution is smaller than the harmonic solution, since both solutions are obtained by taking suprema over classes of functions and the harmonic solution is the supremum over a larger class of functions. Similarly, one can use the Perron method with plurisuperharmonic rather than plurisubharmonic functions to obtain a solution which is plurisuperharmonic and satisfies the Monge–Ampère equation. This solution is greater than the harmonic solution. The Perron method thus yields a plurisubharmonic solution u and a plurisuperharmonic solution v both satisfying the Monge–Ampère equation such that $u \leq h \leq v$, where h is the harmonic solution.

Chapter 10
Uniform Approximation

Abstract The problem of polynomial approximation is introduced along with the appropriate terminology. However, the solution is an open problem. It is not known which domains have the property that holomorphic functions on these domains can be approximated by polynomials.

A fundamental topic in pure and applied mathematics is that of polynomial approximation. Let f be a function defined on a subset $A \subset \mathbb{R}^n$ and suppose f can be uniformly approximated by polynomials arbitrarily well. That is, there is a sequence $\{p_j\}$ of polynomials, which converges uniformly on A to f. Then, the sequence $\{f_j\}$ is uniformly Cauchy on A and so it is also uniformly Cauchy on $\overline{A}$. Consequently, f_j also converges uniformly on $\overline{A}$. We conclude that there is a continuous function on $\overline{A}$, which coincides with f on A. For this reason, if we wish to approximate a function f by polynomials on a set A, it is natural to assume that A is closed and f is continuous on A. Since (nonconstant) polynomials are unbounded on $\mathbb{R}^n$, it is also natural to assume that A is bounded, that is, A is compact. In real analysis, nothing more is required. The Stone–Weierstrass theorem tells us that every continuous function on every compact subset of $\mathbb{R}^n$ can be uniformly approximated by polynomials. In this sense, polynomial approximation is always possible in real analysis.

In complex analysis, the situation is not so simple. We have seen that the uniform limit of holomorphic functions on an open set G in $\mathbb{C}^n$ is holomorphic on G. Consequently, if a continuous function f on a compact subset of $\mathbb{C}^n$ is the uniform limit of (complex) polynomials, then f is necessarily holomorphic on the interior of K.

Since every entire function can be expanded in its Taylor series, it follows that, for a compact set $K \subset \mathbb{C}^n$, approximation by polynomials is equivalent to approximation by entire functions.

Denote by $\mathcal{O}(K)$ the family of functions f holomorphic on a neighborhood of K, where the neighborhood may depend on the function f. Let us say that a compact set $K \subset \mathbb{C}^n$ is a *Runge compactum* if each $f \in \mathcal{O}(K)$ is the uniform limit on K of polynomials. In one complex variable, Runge's theorem states that a compact set

© Springer International Publishing Switzerland 2014
P.M. Gauthier, *Lectures on Several Complex Variables*,
DOI 10.1007/978-3-319-11511-5_10

$K \subset \mathbb{C}$ is a Runge compactum if and only if $\mathbb{C} \setminus K$ is connected. In $\mathbb{C}^n, n > 1$, the characterization of Runge compacta is an open problem and it can be shown that the condition that $\mathbb{C}^n \setminus K$ be connected is neither necessary nor sufficient.

Even better known than the notion of Runge compactum is that of *Runge domain*. A domain $\Omega \subset \mathbb{C}^n$ is called a Runge domain if every $f \in \mathcal{O}(\Omega)$ is the limit of polynomials. When we say that f is the limit of polynomials, we mean in the usual topology for $C(\Omega)$, that of uniform convergence on compact subsets of Ω. Thus, Ω is a Runge domain if and only if, for each $f \in \mathcal{O}(\Omega)$, for each compact $K \subset \Omega$, and for each $\epsilon > 0$, there is a polynomial p such that $|p(z) - f(z)| < \epsilon$, for every $z \in K$.

Problem 36. Every open polydisc Ω in $\mathbb{C}^n$ is a Runge domain.

More generally, it can be shown that every convex domain is a Runge domain. In particular, every open ball is a Runge domain.

It is easy to see that the property of being a Runge domain in $\mathbb{C}^n$ is invariant under automorphisms of $\mathbb{C}^n$. That is, if $\varphi : \mathbb{C}^n \to \mathbb{C}^n$ is a biholomorphic mapping, then it maps Runge domains to Runge domains. However, it is not invariant under biholomorphic mappings of the domains themselves. That is, if Ω is a Runge domain in $\mathbb{C}^n$ and φ maps Ω biholomorphically onto a domain W in $\mathbb{C}^n$, then W need not be a Runge domain. Wermer [19] gives an example of a bounded domain in $\mathbb{C}^2$ which is biholomorphic to the bidisc $\mathbb{D}^2$ but is not a Runge domain.

The property of being a domain of holomorphy is also invariant under automorphisms of $\mathbb{C}^n$. It is natural to give a moment's thought to the relation between Runge domains and domains of holomorphy. A Runge domain need not be a domain of holomorphy and a domain of holomorphy need not be a Runge domain.

For an example in the first direction, the Hartogs figure H is the example par excellence of a domain which is not a domain of holomorphy. However, it is a Runge domain. Indeed, suppose $f \in \mathcal{O}(H)$, K is a compact subset of H, and $\epsilon > 0$. By Theorem 15, f has a holomorphic continuation to the polydisc $\mathbb{D}^2$, which we continue to denote by f. In the polydisc, we may expand f in power series. The partial sums are polynomials which converge uniformly to f on compact subsets of the polydisc and, in particular, on K. Thus, H is a Runge domain.

In the other direction, we have mentioned in the introduction that in $\mathbb{C}$ every domain is a domain of holomorphy, but not all domains are Runge domains. Indeed, Runge's theorem characterizes Runge domains in $\mathbb{C}$ as being precisely those domains whose complement in $\overline{\mathbb{C}}$ is connected.

Chapter 11
Complex Manifolds

Abstract Complex manifolds are introduced as spaces which are locally (complex) Euclidean. They are higher-dimensional analogs of Riemann surfaces.

Complex manifolds are higher dimensional analogs of Riemann surfaces. A manifold is, loosely speaking, a topological space which is locally Euclidean.

Let M be a connected, Hausdorff space having a countable base of open sets. Suppose we are given a covering $\mathcal{U} = \{U_\alpha\}$ of M by open sets and homeomorphisms $\varphi_\alpha : U_\alpha \to V_\alpha$, where each V_α is an open set in real Euclidean space $\mathbb{R}^n$. A pair $(U_\alpha, \varphi_\alpha)$ is called a *chart* and the family of charts $\mathcal{A} = \{(U_\alpha, \varphi_\alpha)\}_\alpha$ is called an *atlas*. The open sets U_α are called *coordinate neighborhoods* and the variable $x^\alpha = \varphi_\alpha(p)$, where $p \in U_\alpha$, is called a *local coordinate* corresponding to U_α.

If $U_\alpha \cap U_\beta \neq \emptyset$, then we have a homeomorphism

$$\varphi_{\alpha\beta} = \varphi_\beta \circ \varphi_\alpha^{-1} : \varphi_\alpha(U_\alpha \cap U_\beta) \to \varphi_\beta(U_\alpha \cap U_\beta).$$

Such a homeomorphism is called a *change of coordinates* for the atlas $\mathcal{A}$. We say that an atlas $\mathcal{A}$ is of *smoothness* k if each change of coordinates $\varphi_{\alpha\beta}$ is of smoothness k.

Two atlases $\mathcal{A}$ and $\mathcal{A}'$ of smoothness k, corresponding, respectively, to coverings $\mathcal{U}$ and $\mathcal{U}'$, are said to be *equivalent* if their union $\mathcal{A} \cup \mathcal{A}'$ is again an atlas of smoothness k. A *real manifold* of smoothness k is a topological space M as above, together with an equivalence class of atlases of smoothness k. The (real) *dimension* of M is the dimension n of the open sets V_α to which the coordinate neighborhoods U_α are homeomorphic. If $k = 0$, we say that M is a *topological manifold*. Loosely speaking, a real manifold is a topological space which is locally real Euclidean.

We shall now introduce *complex manifolds*, which are, loosely speaking, locally complex Euclidean. Indeed, to define a complex manifold of (complex) dimension n, we copy the definition of a real manifold of (real) dimension n. The only difference is that, instead of requiring that the V_α be open sets in real Euclidean space $\mathbb{R}^n$, we require that they be open sets in complex Euclidean space $\mathbb{C}^n$. We may speak of *complex* coordinates, charts, atlases, etc. Thus, a complex manifold of complex dimension n can be considered as a real manifold of real dimension

© Springer International Publishing Switzerland 2014
P.M. Gauthier, *Lectures on Several Complex Variables*,
DOI 10.1007/978-3-319-11511-5_11

$2n$, so it would seem that the study of complex manifolds is merely the study of real manifolds in even real dimensions. If we consider only *topological* manifolds, this point of view is plausible. However, when considering complex manifolds, we usually require a very high level of smoothness. A complex atlas $\mathcal{A}$ is said to be a *holomorphic atlas* if the changes of coordinates $\varphi_{\alpha\beta}$ are biholomorphic. A *holomorphic structure* on M is an equivalence class of holomorphic atlases on M. Often, we shall, as elsewhere in mathematics, merely give a holomorphic atlas $\mathcal{U}$ for a manifold and think of it as the equivalence class of all structures which are (biholomorphically) compatible with it. Of course we shall associate the same holomorphic structure to two holomorphic atlases $\mathcal{U}$ and $\mathcal{V}$ if and only if the two atlases are compatible. Since the union of compatible holomorphic atlases is a holomorphic atlas, for any holomorphic atlas $\mathcal{A}$, there is a maximal holomorphic atlas compatible with $\mathcal{A}$. This is merely the union of all holomorphic atlases compatible with $\mathcal{A}$. Thus, we may think of a holomorphic structure on M as a maximal holomorphic atlas. It seems we have now defined a holomorphic structure on M in three ways: as an equivalence class of holomorphic atlases, as a holomorphic atlas which is maximal with respect to equivalence, or simply as a holomorphic atlas $\mathcal{U}$, meaning the equivalence class of $\mathcal{U}$ or the maximal holomorphic atlas equivalent with $\mathcal{U}$. All that matters at this point is to be able to tell whether two holomorphic structures on M are the same or not. No matter which definition we use, we shall always come up with the same answer. That is, two structures will be considered different with respect to one of the definitions if and only if they are considered different with respect to the other definitions.

Example. Let $M = \mathbb{R}^2 = \{(s,t) : s,t \in \mathbb{R}\}$. We shall consider two atlases $\mathcal{U}$ and $\mathcal{V}$ on M. Each of these atlases will consist of a single chart:

$$\mathcal{U} = \{(\mathbb{R}^2, \varphi)\}, \varphi : \mathbb{R}^2 \to \mathbb{C},$$

where $\varphi(s,t) = z = x + iy$, with $x = s, y = t$ and

$$\mathcal{V} = \{(\mathbb{R}^2, \psi)\}, \psi : \mathbb{R}^2 \to \mathbb{C},$$

where $\psi(s,t) = w = u + iv$, with $u = s, v = -t$. Since the change of charts $z \mapsto w$ is given by $w = \bar{z}$, these two charts are not compatible. Hence the two atlases $\mathcal{U}$ and $\mathcal{V}$ are not compatible, that is, are not equivalent. Thus, the atlases $\mathcal{U}$ and $\mathcal{V}$ define two *different* complex structures on $\mathbb{R}^2$.

A *complex holomorphic manifold* is a topological space M as above, together with a holomorphic structure. We shall consider almost exclusively *holomorphic* complex manifolds. Thus, for brevity, when we speak of a complex structure, we shall mean a holomorphic structure and when we speak of a complex manifold, we shall always (unless otherwise specified) mean a manifold endowed with a complex (holomorphic) structure. A *Riemann surface* is a complex manifold of dimension one. Complex manifolds are higher dimensional analogs of Riemann surfaces. On rare occasions, we shall refer to topological complex manifolds when speaking of complex manifolds whose changes of charts are merely assumed to be homeomorphisms. Later, we shall also define *almost complex* manifolds as real manifolds having a structure which is "almost" complex in a sense to be specified.

Since holomorphy, pluriharmonicity, and plurisubharmonicity of functions are invariant under biholomorphic mappings, these notions may be well defined on complex manifolds. Namely, we define a function f on a complex manifold M to be holomorphic, pluriharmonic, or plurisubharmonic if it is so in each coordinate. More precisely, f is said to be holomorphic, pluriharmonic, or plurisubharmonic on an open set $U \subset M$, if for each coordinate neighborhood U_α which meets U, the composition $f \circ \varphi_\alpha^{-1}$ is, respectively, holomorphic, pluriharmonic, or plurisubharmonic on $\varphi_\alpha(U \cap U_\alpha)$. Similarly, we define a mapping between manifolds to be holomorphic if it is holomorphic in the coordinates. More precisely, a mapping $f : U \to M$ from an open subset U of a complex manifold M of dimension m to a complex manifold N of dimension n is said to be *holomorphic* if, for each chart $(U_\alpha, \varphi_\alpha)$ for which U_α meets U and each chart (V_β, ψ_β) for which V_β meets $f(U \cap U_\alpha)$, the composition $\psi_\beta \circ f \circ \varphi_\alpha^{-1}$ is a holomorphic mapping from the open subset $\varphi_\alpha(f^{-1}(V_\beta) \cap U_\alpha)$ of $\mathbb{C}^m$ into $\mathbb{C}^n$. It is easily verified that holomorphy, pluriharmonicity, and plurisubharmonicity are preserved by holomorphic mappings between manifolds. That is, if g is a holomorphic mapping from an open subset U of a complex manifold M to a complex manifold N and if f is a function defined in a neighborhood of $g(U)$, which is holomorphic, pluriharmonic, or plurisubharmonic, then the composition $f \circ g$ is, respectively, holomorphic, pluriharmonic, or plurisubharmonic on U. It also follows that the composition of holomorphic mappings between manifolds is holomorphic.

The student should verify the following fact. *Let $\mathcal{U}$ and $\mathcal{V}$ be two holomorphic atlases on the same topological manifold M. The atlases $\mathcal{U}$ and $\mathcal{V}$ are compatible if and only if the identity mapping $p \mapsto p$ from the complex manifold $(M, \mathcal{U})$ to the complex manifold $(M, \mathcal{V})$ is biholomorphic.* In other words, two complex structures on the same topological manifold are the same if and only if the identity mapping is biholomorphic with respect to these two complex structures.

Chapter 12
Examples of Manifolds

Abstract Important examples of manifolds are presented, including Lie groups, projective spaces, Grassmann manifolds, and tori.

In this section we give several examples of complex manifolds.

12.1 Domains

Problem 37. Let M be a complex manifold and Ω be a domain in M, that is, an open connected subset. Then, the complex structure of M induces a complex structure on Ω, making Ω a complex manifold. The holomorphic and plurisubharmonic functions on Ω considered as a complex manifold are precisely the holomorphic and plurisubharmonic functions on Ω considered as an open subset of M.

In particular, if Ω is a domain in $\mathbb{C}^n$, then the holomorphic and plurisubharmonic functions on Ω considered as a complex manifold are precisely the holomorphic and plurisubharmonic functions on Ω considered as an open subset of $\mathbb{C}^n$. This shows that complex analysis on manifolds is a generalization of complex analysis on domains in $\mathbb{C}^n$.

12.2 Submanifolds

A connected subset M of $\mathbb{R}^n$ is said to be a *submanifold* of $\mathbb{R}^n$ of smoothness C^ℓ if for each $p \in M$ there is an open neighborhood U_p of p, a number $k \in \{0, \cdots, n\}$, and a C^ℓ-diffeomorphism $f = (f_1, \cdots, f_n)$ of U_p onto an open neighborhood V_0 of the origin such that

$$M \cap U_p = \{t \in U_p : f_{k+1}(t) = \cdots f_n(t) = 0\}.$$

© Springer International Publishing Switzerland 2014
P.M. Gauthier, *Lectures on Several Complex Variables*,
DOI 10.1007/978-3-319-11511-5_12

If we write $s = f(t)$ and $N = f(M \cap U_p)$, then in the local coordinates $s_1, \cdots, s_n$,

$$N \cap V_0 = \{s \in V_0 : s_{k+1} = \cdots s_n = 0\}.$$

The number k is called the *dimension* of the submanifold M at the point p.

We shall say that a subset M of $\mathbb{C}^n$ is a real submanifold of $\mathbb{C}^n$ of dimension k if M is a submanifold of dimension k of the space $\mathbb{R}^{2n}$ underlying $\mathbb{C}^n$.

Analogously, a connected subset M of $\mathbb{C}^n$ is said to be a complex *submanifold* of $\mathbb{C}^n$ if for each $p \in M$ there is an open neighborhood U_p of p, a number $k \in \{0, \cdots, n\}$, and $f = (f_1, \cdots, f_n)$ mapping U_p biholomorphically onto an open neighborhood V_0 of the origin such that

$$M \cap U_p = \{z \in U_p : f_{k+1}(z) = \cdots f_n(z) = 0\}.$$

If we write $\zeta = f(z)$ and $N = f(M \cap U_p)$, then in the holomorphic coordinates $\zeta_1, \cdots, \zeta_n$,

$$N \cap V_0 = \{\zeta \in V_0 : \zeta_{k+1} = \cdots \zeta_n = 0\}.$$

The number k is called the *dimension* of the complex submanifold M at the point p. Obviously, every submanifold of $\mathbb{C}^n$ of dimension k can be thought of as a real submanifold of dimension $2k$ (but not conversely).

Problem 38. Show that a complex submanifold M of $\mathbb{C}^n$ is indeed a complex manifold, and if a function u is holomorphic, pluriharmonic, or plurisubharmonic on the *subset* M, then u is, respectively, holomorphic, pluriharmonic, or plurisubharmonic on the submanifold M.

Submanifolds can be thought of as higher dimensional analogs of curves and surfaces. In physical space considered as $\mathbb{R}^3$, a nonintersecting curve is an example of a one-dimensional real submanifold and a surface is an example of a two-dimensional real submanifold.

We have defined submanifolds of $\mathbb{R}^n$ and $\mathbb{C}^n$. Let Ω be a domain in $\mathbb{R}^n$ or $\mathbb{C}^n$. It is obvious how to define a submanifold of Ω. A submanifold M of a domain Ω is said to be a *closed* submanifold of Ω if M is a closed subset of Ω. For example the open intervals $(0, 1)$ and $(-1, +1)$ are both submanifolds of the open unit disc $\mathbb{D}$. The first is not a closed submanifold of $\mathbb{D}$, whereas the second is. Henceforth, when we speak of a submanifold of a domain Ω, we shall mean a closed submanifold. By a smooth manifold, we shall mean one such that the changes of coordinates are smooth mappings, by which we mean C^1-mappings.

Theorem 50. *Let M be a closed connected subset of a domain Ω in $\mathbb{R}^n$. Then M is a smooth submanifold of Ω if and only if, for each $a \in M$, there exists a neighborhood $U \subset \Omega$ and a smooth mapping $f : U \to \mathbb{R}^m$ such that*

$$U \cap M = \{t \in U : f(t) = 0\}$$

and

$$rank\left(\frac{\partial f}{\partial t}\right) = constant, \quad on \quad U.$$

Proof. Suppose M is a smooth submanifold of Ω and let k be the dimension of M. Fix $a \in M$. From the definition of submanifold, there is a diffeomorphism g of a neighborhood U of a onto an open neighborhood V of the origin in $\mathbb{R}^n$, such that

$$U \cap M = \{t \in U : g_{k+1}(t) = \cdots = g_n(t) = 0\}.$$

Let $\pi : \mathbb{R}^n \to \mathbb{R}^{n-k}$ be the projection $(s_1, \cdots, s_n) \mapsto (s_{k+1}, \cdots, s_n)$. Set $f = \pi \circ f$. Then, f has the properties required of the theorem.

Suppose conversely that M is a closed subset of Ω and f a mapping having the properties stated in the theorem. Then, by the real rank theorem, there are neighborhoods U and V of a and of $0 = f(a)$, polydiscs D^n and D^m in $\mathbb{R}^n$ and $\mathbb{R}^m$ containing the origin, and diffeomorphisms $\varphi : U \to D^n$ and $\psi : V \to D^m$, such that the mapping $\psi \circ f \circ \varphi^{-1}$ has the form $(x_1, \cdots, x_n) \mapsto (x_1, \cdots, x_r, 0, \cdots, 0)$. We may assume that $U \cap M = \{t \in U : f(t) = 0\}$. Thus, if $N = \varphi(M \cap U)$, we have

$$N = \{x \in D^n : \psi \circ f \circ \varphi^{-1}(x) = 0\} = \{x \in D^n : x_1 = \cdots = x_r = 0\}.$$

Thus, the pair (U, φ) is a smooth chart for M at a and M is a smooth submanifold of Ω. $\square$

As an application of the above theorem, let $f : \Omega \to \mathbb{R}$ be a smooth function which is *nonsingular*, that is, $\nabla f(t) \neq 0$, for each $t \in \Omega$. Then, for each $c \in \mathbb{R}$, each component of the level set $f(t) = c$ is a smooth submanifold of Ω. For example, the unit sphere

$$S^{n-1} = \{t \in \mathbb{R}^n : t_1^2 + \cdots + t_n^2 = 1\}$$

is a smooth compact submanifold of dimension $n - 1$ in $\mathbb{R}^n$.

The preceding results on smooth submanifolds of domains in $\mathbb{R}^n$ have analogs for (complex) submanifolds of domains in $\mathbb{C}^n$.

Theorem 51. *Let M be a closed connected subset of a domain Ω in $\mathbb{C}^n$. Then M is a complex submanifold of Ω if and only if, for each $a \in M$, there exists a neighborhood $U \subset \Omega$ and a holomorphic mapping $f : U \to \mathbb{C}^m$ such that*

$$U \cap M = \{z \in U : f(z) = 0\}$$

and

$$rank\left(\frac{\partial f}{\partial z}\right) = constant, \quad on \quad U.$$

Proof. The proof is the same as for the real version using the complex rank theorem rather than the real rank theorem.   □

As an application of the above theorem, let $f : \Omega \to \mathbb{C}$ be a holomorphic function which is nonsingular, that is, $(\partial f/\partial z)(z) \neq 0$, for each $z \in \Omega$. Then, for each $c \in \mathbb{C}$, each component of the level set $f(z) = c$ is a complex submanifold of Ω. For example, each component of the complex sphere

$$\{z \in \mathbb{C}^n : z_1^2 + \cdots + z_n^2 = 1\}$$

is a complex submanifold of dimension $n - 1$ in $\mathbb{C}^n$. Note that the complex sphere is unbounded!

Fermat's last theorem (Wiles' theorem) asserts that the equation

$$x^n + y^n = z^n, \quad n > 2,$$

has no integer solutions with $xyz \neq 0$. Note that, by the above theorem, each component of the intersection of the set

$$\{(x, y, z) \in \mathbb{C}^3 : x^n + y^n - z^n = 0\}$$

with the open set $xyz \neq 0$ is a complex submanifold of the open set $xyz \neq 0$. Wiles' theorem asserts that this submanifold does not intersect any points with integer coordinates.

We have defined the notion of a submanifold of a Euclidean domain. This allows us to define a submanifold of a manifold. We say that a connected subset N of a manifold M is a submanifold of M, if for each chart $\varphi : U \to V$ of M, where V is a Euclidean domain, the set $\varphi(N \cap U)$ is a submanifold of the domain V. It is easy to see that, if M is itself a Euclidean domain, then this coincides with our previous definition of submanifold.

Sometimes, we relax the requirement that a submanifold be connected. In this case, different components may have different dimensions.

12.3   Product Manifolds and Matrices

We leave it to the reader to verify that *the cartesian product of manifolds has a natural manifold structure.*

The set $\mathcal{M}(k, n)$ of (complex) $k \times n$ matrices can be identified with $\mathbb{C}^{kn}$ and so is a complex manifold. If $k \leq n$, the subset $M(k, n)$ of matrices of rank k is an open subset, because a matrix is of rank k if and only if some $k \times k$ minor has nonzero determinant.

We now show that $M(k, n)$ is connected, in fact path-connected. Let A and B be distinct elements of $M(k, n)$, that is, distinct $k \times n$ matrices of rank k. The rows $a_1, \cdots, a_k$ of A are linearly independent and the rows $b_1, \ldots, b_k$ of B are linearly independent. If b_1 is not a multiple of a_1, then a_1 and b_1 span a two-dimensional subspace of $\mathbb{C}^n$. We denote this space by $\mathbb{C}^2$ and its elements by (z, w). For $0 \leq t \leq 1$, let a_t be the vector in $\mathbb{C}^2$ whose (z, w) coordinates are $((1 - t)a_1, t b_1)$. Thus, $a_t; 0 \leq t \leq 1$ is a continuous path in $\mathbb{C}^2 \setminus \{0\}$ from a_1 to b_1. We have constructed $\mathbb{C}^2$ as a subspace of $\mathbb{C}^n$ and so we may consider a_t to be a path in $\mathbb{C}^n$ from a_1 to b_1.

Let A_t be the $k \times n$ matrix whose first row is a_t and whose ith row is a_i for $1 < i \leq k$. Then $A_t; 0 \leq t \leq 1$ is a path in $\mathcal{M}(k, n)$ from A_1 to a matrix whose first row is b_1 and whose subsequent rows are $a_i, 1 < i \leq k$. Call this matrix C_1. Since, for each t, the rows of A_t are linearly independent, A_t is of rank k and so this path is in fact in $M(k, n)$. In a similar fashion, we can construct a path in $M(k, n)$ from C_1 to a matrix C_2 whose first two rows are b_1 and b_2 and whose remaining rows are $a_i; 2 < i \leq k$. After $(k - 1)$ steps, we obtain a path in $M(k, n)$ from C_{k-1} to a matrix all of whose rows are those of B. That is, we arrive at B. By joining these paths together, we have a path in $M(k, n)$ from A to B. Thus $M(k, n)$ is path-connected.

We have shown that $M(k, n)$ is an open and path-connected subset of $\mathcal{M}(k, n)$; in particular it is a subdomain and therefore inherits the manifold structure of $\mathcal{M}(k, n)$.

12.4 Lie Groups

A (complex) *Lie group* G is a (complex) manifold which is also a group such that the group structure is compatible with the manifold structure. That is, the operation of multiplication

$$G \times G \longrightarrow G, \quad (g, h) \mapsto gh$$

and that of inversion

$$G \longrightarrow G, \quad g \mapsto g^{-1}$$

are holomorphic mappings.

As examples of Lie groups, we have the additive groups $\mathbb{C}^n$ and the multiplicative groups $GL(n, \mathbb{C})$ of $n \times n$ invertible matrices. A great deal of mathematical physics is expressed in the language of Lie groups.

12.5   Projective Space

Before introducing projective spaces, we first recall the notion of a quotient topological space.

Let X be a topological space, Y a set, and $f : X \to Y$. The *quotient topology* induced by f is the largest topology on Y such that f is continuous. The open sets in Y for the quotient topology are precisely those sets $V \subset Y$ such that $f^{-1}(V)$ is open.

To each equivalence relation on a set X, we associate the partition of X consisting of equivalence classes. Conversely, to each partition of X, we may associate the equivalence relation defined by saying that two elements of X are equivalent if they belong to the same member of the partition. This gives a one-to-one correspondence between equivalence relations $\sim$ on X and partitions $\mathcal{P}$ of X. A *quotient set* of X is defined as a set $X/\sim$ of equivalence classes with respect to an equivalence relation $\sim$ on X. There is a natural projection of X onto a quotient set $X/\sim$ denoted $p : X \to X/\sim$ defined by sending a point x to its equivalence class $[x]$. Let us say that a function f on X is $\sim$ *invariant* if $f(x) = f(y)$, whenever $x \sim y$. The projection induces a natural bijection between $\sim$-invariant functions on X and functions on $X/\sim$.

Let X be a topological space and $\sim$ an equivalence relation on X. The *quotient topological space* induced by an equivalence relation $\sim$ on X is the quotient set $X/\sim$ endowed with the quotient topology induced by the natural projection $p : X \to X/\sim$. We sometimes speak of the quotient topology induced by an equivalence relation (partition) as the *identification topology*, since we identify points in the same equivalence class (member of the partition).

As an example, let X be the closed unit interval $[0, 1]$ with the usual topology and let $\mathcal{P}$ be the partition which identifies 0 and 1. That is, the members of the partition are the set $\{0, 1\}$ and the singletons $\{t\}, 0 < t < 1$. The quotient space $[0, 1]/\mathcal{P}$ is then the circle with its usual topology.

If a topological space X is connected and has a countable base for its topology, then any quotient space of X is also connected and has a countable base for its topology. However, a quotient space of a Hausdorff space need not be Hausdorff.

Lemma 52. *Let $X/\sim$ be a quotient space of a Hausdorff space X with respect to an equivalence relation $\sim$ on X. Then, $X/\sim$ is also Hausdorff if and only if for each $[x] \neq [y]$ in $X/\sim$, there are disjoint open subsets U and V of X, both of which are unions of equivalence classes, such that $[x] \subset U$ and $[y] \subset V$.*

There is a general notion of a quotient manifold, which we shall not define in this section. We do present, however, the most important example, complex projective space $\mathbb{P}^n$ of dimension n, which we think of as a compactification of the complex Euclidean space $\mathbb{C}^n$ obtained by adding "points at infinity" to $\mathbb{C}^n$. For $n = 1$ we obtain the Riemann sphere $\mathbb{P}^1 = \overline{\mathbb{C}}$. Projective space $\mathbb{P}^n$ is the most fundamental space for algebraic geometry.

We define *projective space* $\mathbb{P}^n$ as the set of all complex lines in $\mathbb{C}^{n+1}$ which pass through the origin. Let us denote a point $\omega \neq 0$ in $\mathbb{C}^{n+1}$ by $(\omega_0, \cdots, \omega_n)$. Two points ω and ω' both different from zero lie on the same line through the origin if and only if $\omega = \lambda \omega'$ for some $\lambda \in \mathbb{C}$. This is an equivalence relation $\sim$ on $\mathbb{C}^{n+1} \setminus \{0\}$. Projective space is the quotient space

$$\mathbb{C}^{n+1} \setminus \{0\}/ \sim .$$

Since $\mathbb{C}^{n+1} \setminus \{0\}$ is connected and has a countable base for its topology, the same is true of the quotient space $\mathbb{P}^n$.

Let us show that $\mathbb{P}^n$ is Hausdorff. We invoke Lemma 52. Suppose then that $[\omega]$ and $[\omega']$ represent two distinct complex lines in $\mathbb{C}^{n+1}$ passing through the origin. For any subset E of the unit sphere S in $\mathbb{C}^{n+1}$, we denote

$$C(E) = \{e^{i\theta} w : w \in E, \theta \in [0, 2\pi]\}.$$

We may assume that the points ω and ω' lie on the sphere S. Let u_j and v_j be sequences of open subsets of S decreasing, respectively, to ω and ω'. Suppose, for $j = 1, \cdots$, there is a point $p_j \in C(u_j) \cap C(v_j)$. Since this sequence lies on the sphere S which is compact, we may assume that the sequence converges. The limit point must lie on both of the circles $C(\omega)$ and $C(\omega')$, which however are disjoint. This contradiction shows that there exist open neighborhoods u and v of ω and ω', respectively, in S such that $C(u) \cap C(v) = \emptyset$. Set

$$U = \{[w] : w \in C(u)\}, \quad V = \{[w] : w \in C(v)\}.$$

Then, U and V are disjoint open subsets of $\mathbb{C}^{n+1} \setminus \{0\}$ which contain $[\omega]$ and $[\omega']$, respectively, and which are both unions of equivalence classes. By Lemma 52, the quotient space $\mathbb{P}^n$ is Hausdorff.

Let us denote the equivalence class (the line passing through ω) by the "homogeneous coordinates" $[\omega] = [\omega_0, \cdots, \omega_n]$. Let

$$U_j = \{[\omega_0, \cdots, \omega_n] : \omega_j \neq 0\}, \quad j = 0, \cdots, n,$$

and define a mapping $\varphi_j : U_j \to \mathbb{C}^n$ by

$$\varphi_j([\omega_0, \cdots, \omega_n]) = \left(\frac{\omega_0}{\omega_j}, \cdots, \frac{\omega_{j-1}}{\omega_j}, \frac{\omega_{j+1}}{\omega_j} \cdots, \frac{\omega_n}{\omega_j} \right).$$

The family $U_j, j = 0, \cdots, n$ is a finite cover of $\mathbb{P}^n$ by open sets and the mappings φ_j are homeomorphisms from U_j onto $\mathbb{C}^n$. Thus, $\mathbb{P}^n$ has a countable base, since it has a finite cover by open sets, each of which has a countable base. We have shown that $\mathbb{P}^n$ *is a connected Hausdorff space whose topology has a countable base* and we have exhibited a topological atlas $\mathcal{A} = \{(U_j, \varphi_j) : j = 0, \cdots, n\}$. Thus, $\mathbb{P}^n$ is a topological manifold of complex dimension n.

Problem 39. The atlas $\mathcal{A} = \{(U_j, \varphi_j) : j = 0, \cdots, n\}$ is a *holomorphic* atlas giving projective space $\mathbb{P}^n$ the structure of a complex manifold.

We may express projective space $\mathbb{P}^n$ as the disjoint union of U_0 which is biholomorphic to $\mathbb{C}^n$ and the set $\{[\omega] = [0, \omega_1, \cdots, \omega_n] : \omega \neq 0\}$ which is in one-to-one correspondence with the points of $\mathbb{P}^{n-1}$ in homogeneous coordinates. Thus,

$$\mathbb{P}^n = \mathbb{C}^n \cup \mathbb{P}^{n-1}.$$

Consequently, we may think of projective space as a compactification of Euclidean space obtained by adding "points at infinity."

In view of our definition of projective space, it is natural to define $\mathbb{P}^0$ to be the space of complex lines through the origin in $\mathbb{C}$. Thus, $\mathbb{P}^0$ is a singleton which we may think of as a zero-dimensional complex manifold. Let us denote this ideal point by ∞. The preceding formula in this case becomes

$$\mathbb{P} = \mathbb{C} \cup \{\infty\},$$

the one-point compactification of $\mathbb{C}$. The complex projective space of dimension one is therefore the Riemann sphere.

12.6 Grassmann Manifolds

For $k \leq n$, let $Gr(k, n)$ denote the family of k-dimensional vector subspaces of $\mathbb{C}^n$. There is a natural way of endowing $Gr(k, n)$ with the structure of a complex manifold. These manifolds are called Grassmann manifolds. In particular, $Gr(1, n + 1)$ is projective space $\mathbb{P}^n$.

Denote by $\mathcal{M}(k, n)$ the set of $k \times n$ complex matrices. We identify $\mathcal{M}(k, n)$ with $(\mathbb{C}^n)^k$ by considering the rows $a_1, \cdots, a_k$ of a matrix $A \in \mathcal{M}(k, n)$ as vectors in $\mathbb{C}^n$. Denote by $M(k, n)$ the subset of $\mathcal{M}(n, k)$ consisting of matrices of rank k. Since a matrix is of rank k if and only if some $k \times k$ minor has determinant different from zero, it follows that $M(k, n)$ is open in $\mathcal{M}(n, k)$. Thus, $M(k, n)$ inherits the complex structure of $\mathcal{M}(k, n)$. We now define a mapping

$$\pi : M(k, n) \longrightarrow Gr(k, n)$$

as follows. For $A \in M(k, n)$ we put πA equal to the k-subspace of $\mathbb{C}^n$ spanned by the rows of A. Denote, as usual, the group of invertible $k \times k$ complex matrices by $GL(k) = GL(k, \mathbb{C})$. The group $GL(k)$ acts on $\mathcal{M}(k, n)$ in the sense that, for $g \in GL(k)$ and $A \in \mathcal{M}(k, n)$, the product gA is a matrix in $\mathcal{M}(k, n)$. The mapping π is invariant under this action. That is, $\pi(gA) = \pi A$, for $A \in M(k, n)$, since multiplying A by g amounts to a change of basis for the k-subspace associated to A. Indeed, if $a_1, \cdots, a_k$ are the rows of A, then the rows of gA are $b_1, \cdots, b_k$, where, for $j = 1, \cdots, k$, we have $b_j = g_{j1}a_1 + \cdots + g_{jk}a_k$. Thus each row of gA is a linear combination of the rows of A. Denote by $M(k,n)/GL(k)$ the set of orbits $[A]$ under the action of $GL(k)$. That is

$$[A] = GL(k)A = \{gA : g \in GL(k)\}.$$

We have shown that

$$M(k,n)/GL(k) \equiv Gr(k,n),$$

and we give $Gr(k,n)$ the quotient topology induced by the mapping π.

To show that $Gr(k,n)$ is Hausdorff, we need to show that if $[A] \neq [B]$, then there are neighborhoods U and V of $[A]$ and $[B]$, respectively, such that $\pi U \cap \pi V = \emptyset$. It is sufficient to construct a continuous function $\rho : Gr(k,n) \to \mathbb{R}$ which separates $[A]$ and $[B]$. For $w \in \mathbb{C}^n \setminus \{0\}$, and $X \in Gr(k,n)$, set $\rho_w(X) = d(w/|w|, X)$, where d denotes distance in $\mathbb{C}^n$.

For fixed w, the mapping

$$M(k,n) \longrightarrow Gr(k,n) \longrightarrow \mathbb{R}$$

$$A \longmapsto \pi(A) \longmapsto \rho_w(\pi(A)),$$

given by $\rho_w \circ \pi$ is continuous and since $Gr(k,n)$ has the quotient topology induced by π, it follows that ρ_w is also continuous. Now, suppose $[A] \neq [B]$ and choose $w \in \pi(A) \setminus \pi(B)$. Then, $\rho_w([A]) = 0$ while $\rho_w([B]) \neq 0$. Thus, the continuous function ρ_w separates $[A]$ and $[B]$. Therefore, $[A]$ and $[B]$ have disjoint neighborhoods, so $Gr(k,n)$ is Hausdorff.

Let us define charts on $Gr(k,n)$. We shall start by covering $Gr(k,n)$ by finitely many open sets (future charts), in a similar fashion as for projective space. Let J be a subset of k indices among $\{1, \cdots, n\}$. Let $e_1, \cdots, e_n$ be the standard basis vectors of $\mathbb{C}^n$ and let C_J be the $(n-k)$-dimensional subspace of $\mathbb{C}^n$ spanned by the vectors $e_j, j \notin J$. Define the open subset U_J of $Gr(k,n)$ as follows:

$$U_J = \{X \in Gr(k,n) : X \cap C_J = \{0\}\}.$$

We wish to define a coordinate map from U_J to a complex Euclidean space. That is, we wish to assign coordinates to each $X \in U_J$. By acting on X with elements $g \in GL(k)$, we can find a representative gX for which the $k \times k$ minor X_J determined by J is the identity matrix I_k. This representation is unique. Indeed, g also acts on X_J and $gX_J = I_k$. Thus, $g = X_J^{-1}$ so g is uniquely determined and consequently also the representative gX. The $(k \times n)$-matrix gX has I_k as minor X_J, and we assign as coordinates of X the other $k(n-k)$ entries. This gives a map $\phi_J : U_J \to \mathbb{C}^{k(n-k)}$. For an arbitrary choice of $k(n-k)$ complex numbers, there is an $X \in U_J$, for which ϕ_J assigns these values. Thus, ϕ is a surjection. It is also an injection, for if g_1X_1

and $g_2 X_2$ both have I_k as J-minor and agree at other terms, then $g_1 X_1 = g_2 X_2$, so $X_1 = g_1^{-1} X_2$. That is, X_1 and X_2 are equivalent. We have that $\phi_J : U_J \to \mathbb{C}^{k(n-k)}$ is a bijection.

We can arrive at the mapping ϕ_J in a different way. Let V_J be the subset of $M(k, n)$ consisting of matrices having I_k as $(k \times k)$ minor determined by J. Then, π maps V_J homeomorphically onto U_J. Now, viewing $M(k, n)$ as an open subset of $\mathcal{M}(k, n) \equiv (\mathbb{C}^n)^k$, let W_J be the subset of $(\mathbb{C}^n)^k$ corresponding to V_J. There is a linear automorphism L_J representing $(\mathbb{C}^n)^k$ as $\mathbb{C}^{k^2} \times \mathbb{C}^{k(n-k)}$, for which W_J corresponds to $\mathbb{C}^{k(n-k)}$. Let $P_J : \mathbb{C}^{k^2} \times \mathbb{C}^{k(n-k)} \to \mathbb{C}^{k(n-k)}$ be the projection and let $\Psi_J : (\mathbb{C}^n)^k \to \mathbb{C}^{k(n-k)}$ be the holomorphic surjective mapping $\Psi_J = P_J \circ L_J$. Let F_J be the restriction of Ψ_J to the open set $W(k, n) \subset (\mathbb{C}^n)^k$. Identifying W_J and V_j, we may consider F_J as a holomorphic mapping $W(k, n) \to \mathbb{C}^{k(n-k)}$, whose restriction Φ_J to V_J is a homeomorphism of V_j onto $\mathbb{C}^{k(n-k)}$. By construction, $\Phi_J = \phi_J \circ \pi$, and since U_J has the quotient topology induced by V_J and π, it follows that ϕ_J is also a homeomorphism. As J varies over all subsets of k elements of $\{1, \cdots, n\}$, we obtain an open cover of $Gr(k, n)$ by such U_J and since the $\phi_J : U_J \to \mathbb{C}^{k(n-k)}$ are homeomorphisms, we have that $Gr(k, n)$ is a complex topological manifold of dimension $k(n-k)$. By construction, it is not hard to see that the change of coordinates on the intersection of two coordinate charts in $\mathbb{C}^{k(n-k)}$ is holomorphic. Thus, $G(k, n)$ is not only a topological complex manifold but, indeed, a (holomorphic) complex manifold.

12.7 Tori

In this section, we shall present, as an example of a complex manifold, the complex n-torus. But first we present the real n-torus.

In $\mathbb{R}^n$, let $\omega_1, \cdots, \omega_n$ be linearly independent. Let L be the lattice generated by these vectors:

$$L = \{k_1 \omega_1 + \cdots + k_n \omega_n : k_j \in \mathbb{Z}\} = \mathbb{Z}\omega_1 + \cdots + \mathbb{Z}\omega_n.$$

Two points x and y in $\mathbb{R}^n$ are said to be equivalent mod L if and only if $y = x + \omega$ for some $\omega \in L$. The *real n-torus* induced by L is the quotient space with respect to this equivalence relation and we denote it by $\mathbb{R}^n/L$. If L' is the lattice on $\mathbb{R}^n$ determined by another set of independent vectors $\omega_1', \cdots, \omega_n'$, let $f : \mathbb{R}^n \to \mathbb{R}^n$ be a linear change of basis in $\mathbb{R}^n$ mapping the basis $\omega_1, \cdots, \omega_n$ to the basis $\omega_1', \cdots, \omega_n'$. This is a homeomorphism which maps the lattice L to the lattice L' and two vectors are equivalent mod L if and only if their images are equivalent mod L'. Thus, f induces a homeomorphism of the quotient spaces $\mathbb{R}^n/L$ and $\mathbb{R}^n/L'$. All real n-tori are thus homeomorphic to the standard real n-torus arising from the standard basis $e_1, \cdots, e_n$ of $\mathbb{R}^n$.

It can be shown that *any real n-torus is homeomorphic to* $(S^1)^n$, *the n-fold product of the unit circle* $S^1 = \{x \in \mathbb{R}^2 : |x| = 1\}$.

Problem 40. Verify this for the case $n = 1$.

Clearly, $(S^1)^n$ is a compact real manifold of dimension n and so the same is true of any real n-torus.

We shall now construct complex tori. In $\mathbb{C}^n$, let $\omega_1, \cdots, \omega_{2n}$ be $\mathbb{R}$-independent and let L be the associated lattice:

$$L = \{k_1\omega_1 + \cdots + k_{2n}\omega_{2n} : k_j \in \mathbb{Z}\} = \mathbb{Z}\omega_1 + \cdots + \mathbb{Z}\omega_{2n}.$$

Two points z and ζ in $\mathbb{C}^n$ are said to be *equivalent mod L* if and only if $\zeta = z + \omega$ for some $\omega \in L$. The *complex n-torus* induced by L is the quotient space with respect to this equivalence relation and we denote it by $\mathbb{C}^n/L$. If we think of the real Euclidean space $\mathbb{R}^{2n}$ underlying the complex Euclidean space $\mathbb{C}^n$, then we see that the complex n-torus $\mathbb{C}^n/L$ can be (topologically) identified with the real $2n$-torus $\mathbb{R}^{2n}/L$ and hence with $(S^1)^{2n}$. Thus, the complex n-torus is a compact real manifold of (real) dimension $2n$. We shall endow the complex n-torus with a complex structure with respect to which it is a complex manifold of (complex) dimension n.

For $z \in \mathbb{C}^n$, let $B(z, r)$ be the ball of center z and radius r and set

$$[B(z, r)] = \bigcup_{\zeta \sim z} B(\zeta, r) = \bigcup_{w \in B(z,r)} [w]$$

$$U([z], r) = \{[w] : w \in B(z, r)\}.$$

By abuse of notation, $[w]$ denotes the equivalence class of w as subset of $\mathbb{C}^n$ in the first equation and denotes the corresponding point of $\mathbb{C}^n/L$ in the second equation. If p is the natural projection from $\mathbb{C}^n$ onto $\mathbb{C}^n/L$, then $p([B(z, r)]) = U([z], r)$. Thus, $U([z], r)$ is an open neighborhood of the point $[z]$ in the complex n-torus $\mathbb{C}^n/L$.

We claim that $|\omega|$ is bounded below for $\omega \in L \setminus \{0\}$. Consider first the lattice L_0 generated by the standard basis $e_1, \cdots, e_{2n}$ of the underlying real vector space $\mathbb{R}^{2n}$. If $\omega \in L_0$, then $\omega = k_1 e_1 + \cdots + k_{2n} e_{2n}$. Thus,

$$\min\{|\omega| : \omega \in L_0, \omega \neq 0\} = 1. \tag{12.1}$$

Now the $\mathbb{R}$-linearly independent vectors $\omega_1, \cdots, \omega_{2n}$ generating the lattice L can be obtained from the standard basis $e_1, \cdots, e_{2n}$ by isomorphism of $\mathbb{R}^{2n}$ and this isomorphism also maps the lattice L_0 to the lattice L. Since this isomorphism is bilipschitz, it follows from (12.1) that for some $r_L > 0$,

$$\inf\{|\omega| : \omega \in L, \omega \neq 0\} \geq r_L. \tag{12.2}$$

From (12.2) we see that if $|z - \zeta| < r_L$, then z and ζ are not equivalent. From (12.2) we also see that $[z]$ is a discrete set, since $|a - b| \geq r_L$, for any two distinct points $a, b \in [z]$. Thus, if ζ is not equivalent to z, it follows that ζ is at a positive distance from $[z]$, since ζ is not in the discrete set $[z]$. Further, we claim that if $[z] \neq [\zeta]$, then these two sets are at a positive distance from one another. Suppose not. Then, there are $z_j \sim z$ and $\zeta_j \sim \zeta$ with $|z_j - \zeta_j| \to 0$. We have $z_j = z + \alpha_j$ for some $\alpha_j \in L$. Thus, $\zeta_j - \alpha_j$ is a sequence of points in $[\zeta]$ which converges to z. This contradicts the fact that $[\zeta]$ is at a positive distance from z. We have established that if $[z] \neq [\zeta]$, then $[z]$ and $[\zeta]$ are at a positive distance $2r > 0$ from each other. Thus, the open sets $[B(z, r)]$ and $[B(\zeta, r)]$ in $\mathbb{C}^n$ are disjoint. It follows that *the complex n-torus is Hausdorff.*

Let us now show that the complex n-torus $\mathbb{C}^n / L$ carries a natural structure of a complex manifold of dimension n which is induced by the projection. Notice that the distance between any two points in the same equivalence class $[z]$ is bounded below by r_L. Choose $r < r_L/2$ and for each point $[z] \in \mathbb{C}^n / L$, denote by V_z the open ball $B(z, r)$ in $\mathbb{C}^n$. Set

$$U_{[z]} = p(V_z) \quad \text{and} \quad \varphi_{[z]} = (p \mid_{V_z})^{-1}.$$

Then $U_{[z]}$ is an open neighborhood of the point $[z]$ in the complex n-torus $\mathbb{C}^n / L$. The family $\{(U_{[z]}, \varphi_{[z]})\}$ is an atlas for the complex n-torus $\mathbb{C}^n / L$. Indeed, the projection is both open and continuous from the definition of the quotient topology. Moreover, no two points in $V_z = B(z, r)$ are equivalent, since $2r < r_L$. Thus, p restricted to V_z is a homeomorphism of V_z onto $U_{[z]}$. The changes of coordinates are holomorphic since $\varphi_{[z]} \circ \varphi_{[\zeta]}^{-1}$ is the identity if $V_z \cap V_\zeta \neq \emptyset$. Thus, *the complex n-torus $\mathbb{C}^n / L$ is a compact complex manifold of dimension n.*

12.8 The Quotient Manifold with Respect to an Automorphism Group

We have considered several specific instances of manifolds obtained as quotients of other manifolds, namely projective space, tori, and Grassmannians. In this section, we describe a general class of quotient manifolds. Let M be a complex manifold. We denote by $Aut(M)$ the group of biholomorphic mappings of M onto itself. $Aut(M)$ is called *the automorphism group of M.* Obviously, the automorphism group $Aut(M)$ is a *biholomorphic invariant.* That is, if M and N are biholomorphic complex manifolds, then their automorphism groups $Aut(M)$ and $Aut(N)$ are isomorphic.

The simplest (and most important) example of a complex manifold is $\mathbb{C}^n$. For $n = 1$, the automorphism group $Aut(\mathbb{C})$ is easy to describe. Indeed, it is easy to see that the biholomorphic mappings $f : \mathbb{C} \to \mathbb{C}$ are precisely the invertible affine transformations $z \mapsto az + b, a, b \in \mathbb{C}, a \neq 0$.

For every $n = 1, 2, \cdots$, the affine transformations $z \mapsto Az + b$, where $z \in \mathbb{C}^n$, A is $n \times n$ invertible matrix, and $b \in \mathbb{C}^n$, are obviously in $Aut(\mathbb{C}^n)$. However, for $n > 1$, the automorphism group $Aut(\mathbb{C}^n)$ contains biholomorphic mappings which are not of this form. In fact $Aut(\mathbb{C}^n)$ is huge and very complicated, for $n > 1$. For example, for each entire function f, the *shear* mapping $\mathbb{C}^2 \to \mathbb{C}^2$ defined by $(z, w) \mapsto (z, f(z) + w)$ is biholomorphic. To check this, it is trivial to verify that this holomorphic mapping is a bijection. The complex Jacobian matrix is invertible, so by the inverse mapping theorem, the inverse mapping is also holomorphic. More generally, mappings $(z, w) \mapsto (z, f(z) + h(z)w)$, where h is an entire function having no zeros, are also automorphisms of $\mathbb{C}^2$ called *overshears*. This is proved in the same way as for shears. We see that, although $Aut(\mathbb{C})$ is a one-dimensional complex vector space, $Aut(\mathbb{C}^2)$ contains an infinite-dimensional complex vector space.

We mentioned in the introduction that a striking event at the beginning of the study of several complex variables was the discovery by Poincaré that the automorphism groups $Aut(\mathbb{B}^2)$ and $Aut(\mathbb{D}^2)$ of the ball and the bidisc in $\mathbb{C}^2$ are not isomorphic and hence the ball and the bidisc are not biholomorphic. For a different approach to this problem, we recommend the excellent monograph by Steven G. Krantz [9].

It is easy to see that *automorphisms are proper mappings*. Indeed, suppose M is a complex manifold and $f \in Aut(M)$. Suppose that f is not proper. Then, there exists a compact set $K \subset M$ such that $f^{-1}(K)$ is not compact. Thus, there is a sequence $z_j \in f^{-1}(K)$ with no convergent subsequence. However, there is a subsequence $z_{j(k)}$ such that $f(z_{j(k)})$ converges to some point $b \in K$. Set $a = f^{-1}(b)$. Since f^{-1} is continuous, $f^{-1}(f(z_{j,k})) = z_{j,k}$ converges to a, which contradicts the choice of the sequence z_j. Thus, f is indeed proper.

It follows from the inverse mapping theorem that if $f \in Aut(\mathbb{C}^n)$, then $\det J_f(z) \neq 0$ for each z. The converse is not true even in one variable as the exponential function shows. Perhaps the most famous conjecture in algebraic geometry is that the converse is true for polynomial automorphisms.

Jacobian Conjecture If $p : \mathbb{C}^n \to \mathbb{C}^n$ is a polynomial mapping and $\det J_p(z) \neq 0$, for each $z \in \mathbb{C}^n$, then $p \in Aut(\mathbb{C}^n)$.

Of course it is true for $n = 1$. Further, for any n, we can show that *if p satisfies the hypotheses of the Jacobian Conjecture, then $\det J_p$ is a constant.* Indeed, suppose $\det J_p(z) \neq 0$ for each $z \in \mathbb{C}^n$. We write $P(z) = \det J_p(z)$ and note that P is a polynomial $P(z) = P(z_1, \cdots, z_n)$. If we allow some coordinate z_j to vary but fix the other $n - 1$ coordinates, we obtain a polynomial P_j of the single variable z_j which has no zeros. By the Fundamental Theorem of Algebra, P_j is constant. We have shown that the restriction of the polynomial P to any complex line in a coordinate direction is constant. Since any two points of $\mathbb{C}^n$ can be joined by a path employing such complex lines, these constants are the same. That is, $P = \det J_p$ is constant as asserted.

There have been several incorrect papers claiming to confirm the Jacobian Conjecture in dimension 2. For a partial result in dimension 2, see for example [6].

If G is a subgroup of $Aut(M)$ and $p \in M$, the G-*orbit* of p is $[p] = gp : g \in G$. Let M/G be the space of G-orbits with the quotient topology induced by the projection

$$M \to M/G$$
$$p \mapsto [p] = Gp.$$

A group G of automorphisms of M acts *properly discontinuously* if for each compact $K \subset M$,

$$g(K) \cap K \neq \emptyset,$$

for at most finitely many $g \in G$.

Lemma 53. *If G acts properly discontinuously on M, then M/G is Hausdorff.*

Proof. Let $p_1 \neq p_2$ in M and let (φ_1, U_1) and (φ_2, U_2) be disjoint charts for p_1 and p_2. We may suppose that $\varphi_j(U_j) \supset (|z| \leq 1)$ and $\varphi_j(p_j) = 0$. Set $K_j = \varphi_j^{-1}(|z| \leq 1)$ and

$$A_n = \{p \in U_1 : |\varphi_1(p)| < 1/n\}, \quad B_n = \{p \in U_2 : |\varphi_2(p)| < 1/n\}.$$

Denote by π the projection of M onto M/G. Suppose $\pi(A_n) \cap \pi(B_n) \neq \emptyset$ for each n. Then, there is a $b_n \in B_n$ such that

$$b_n \in \pi^{-1}\pi A_n = \bigcup_{g \in G} g(A_n).$$

Thus, there is a $g_n \in G$ and a point $a_n \in A_n$ such that $g_n(a_n) = b_n$. Thus, $g_n(K) \cap K \neq \emptyset$, where $K = K_1 \cup K_2$. Since G is properly discontinuous, $\{g_1, g_2, \cdots\}$ is finite. For a subsequence, which we continue to denote $\{g_n\}$, we have $g_n = g$, $n = 1, 2, \cdots$:

$$g(p_1) = \lim g(a_n) = \lim g_n(a_n) = p_2.$$

Thus, p_1 and p_2 are G-equivalent.

To show that M/G is Hausdorff, suppose $[p_1]$ and $[p_2]$ are distinct. Thus, by the previous paragraph, there is an n such that $\pi(A_n) \cap \pi(B_n) = \emptyset$. Since π is open, $\pi(A_n)$ and $\pi(B_n)$ are disjoint open neighborhoods of $[p_1]$ and $[p_2]$. Thus, M/G is Hausdorff.   □

A group $G \subset Aut(M)$ is said to act *freely* on M if the only $g \in G$ having a fixed point is id_M.

Theorem 54. *Let M be a complex manifold and G a subgroup of $Aut(M)$. If G acts freely and properly discontinuously on M, then the projection $\pi : M \to M/G$ induces on M/G the structure of a complex manifold and the projection is holomorphic.*

Proof. The quotient space M/G is connected and second countable and from the lemma it is Hausdorff. We need only show that π induces a holomorphic atlas.

Fix $p \in M$; let (φ_p, U) be a chart at p such that $\varphi_p(U) \supset (|z| \leq 1)$ and $\varphi_p(p) = 0$. Set $K = \varphi_p^{-1}(|z| \leq 1)$ and define A_n as in the proof of the lemma. We show that $\pi|_{A_n}$ is injective for sufficiently large n. Suppose not. Then, there are points $a_n \neq b_n$ in A_n with $\pi(a_n) = \pi(b_n)$. Hence, for some $g_n \in G$, we have $g_n(a_n) = b_n$. Since $a_n \neq b_n$, $g_n \neq id_M$. We have $b_n \in K \cap g_n(K)$. Since G is properly discontinuous, it follows that $\{g_1, g_2, \cdots\}$ is finite. Some subsequence which we continue to denote $\{g_n\}$ is constant, $g_n = g$. We continue to denote the corresponding subsequences of $\{a_n\}$ and $\{b_n\}$ as $\{a_n\}$ and $\{b_n\}$, respectively. Then $g(a_n) = b_n$. Now $a_n, b_n \to p$ and so by continuity, $g(p) = p$. Since G acts freely and $g \neq id_M$, this is a contradiction. Hence, for each p, there is an A_n which we denote by A_p on which π is a homeomorphism. Thus, M/G is a manifold if we take as charts the family $(\varphi_p \circ \pi\,|_{A_p}^{-1}, \pi(A_p))$, $p \in M$. It is easy to check that the change of charts is biholomorphic so that M/G is in fact a complex manifold. $\square$

Example. The complex n-torus $\mathbb{C}^n/L$ is an example of a quotient manifold of this type. Recall that L is a lattice

$$L = \{k_1\omega_1 + \cdots + k_{2n}\omega_{2n} : k_j \in \mathbb{Z}\} = \mathbb{Z}\omega_1 + \cdots + \mathbb{Z}\omega_{2n},$$

where $\omega_1, \cdots, \omega_{2n}$ are $\mathbb{R}$-linearly independent vectors in $\mathbb{C}^n$. For $a \in \mathbb{C}^n$, denote by τ_a the translation automorphism $\tau_a(z) = z + a$. Let

$$G_L = \langle \tau_{\omega_1}, \cdots, \tau_{\omega_{2n}} \rangle$$

be the subgroup of $Aut(\mathbb{C}^n)$ generated by the translations $\tau_1, \cdots, \tau_{2n}$.

Write $k = (k_1, \cdots, k_{2n})$ and let g_k be the automorphism of $\mathbb{C}^n$ given by

$$z \mapsto g_k(z) = z + k_1\omega_1 + \cdots + k_{2n}\omega_{2n}.$$

Then $G_L = \{g_k : k \in \mathbb{Z}^n\}$. Suppose $g_k \in G_L$ has a fixed point $z \in \mathbb{C}^n$, so $g_k(z) = z$. That is,

$$z + (k_1\omega_1 + \cdots + k_{2n}\omega_{2n}) = z$$

and consequently

$$k_1\omega_1 + \cdots + k_{2n}\omega_{2n} = 0.$$

Since $\omega_1, \cdots, \omega_{2n}$ are $\mathbb{R}$-linearly independent, $k = 0$. That is, $g_k = id$. We have shown that the only element in G_L having a fixed point is the id. Thus, the group G_L acts freely on $\mathbb{C}^n$.

Let us verify that G_L acts properly discontinuously on $\mathbb{C}^n$. Let $K \subset \mathbb{C}^n$. Since K is compact, there is a $M > 0$ such that $|z| < M$ for all $z \in K$ and by formula (12.2) we have $r = \min |\omega_j| > 0$. Suppose $g_k(K) \cap K \neq \emptyset$. Then, there is a $z \in K$ with $|g_k(z)| \leq M$:

$$M \geq |z + \sum k_j \omega_j| \geq |\sum k_j \omega_j| - |z| \geq \max |k_j \omega_j| - |z| \geq r \cdot \max |k_j| - M.$$

Thus $\max |k_j| \leq 2M/r$. Clearly, only finitely many k satisfy this inequality. That is, there are only finitely many g_k for which some point $z \in K$ has the property that $g_k(z) \in K$. We have shown that G_L acts properly discontinuously on $\mathbb{C}^n$.

By Theorem 54, $\mathbb{C}^n / G_L$ is a complex manifold and the projection $\pi : \mathbb{C}^n \to \mathbb{C}^n / G_L$ is holomorphic.

The complex torus $\mathbb{C}^n / L$ and the complex manifold $\mathbb{C}^n / G_L$ are both defined as quotient spaces of $\mathbb{C}^n$. In the first case $z \sim \zeta$ if $\zeta = z + \omega$, $\omega \in L$. In the second case $z \sim \zeta$ if $\zeta = g_k(z)$, $k \in \mathbb{Z}^n$. Since L is defined as the set of ω of the form $k_1 \omega_1 + \cdots + k_{2n} \omega_{2n}$ and $g_k(z)$ is defined as $z + (k_1 \omega_1 + \cdots + k_{2n} \omega_{2n})$, we see that the two equivalence relations are the same. That is, the two quotient spaces are the same. We have shown that *the manifold $\mathbb{C}^n / G_L$ induced by the subgroup G_L of Aut $(\mathbb{C}^n)$ is the complex torus $\mathbb{C}^n / L$ induced by the lattice L.*

12.9 Spread Manifolds

In this subsection we introduce the notion of a manifold N *spread* (or *étalé*) over a manifold M.

Suppose X and Y are topological spaces and let $\varphi : Y \longrightarrow X$. We shall say that Y is *spread over* X by φ if φ is a local homeomorphism, i.e., each $y \in Y$ has an open neighborhood V_y so that the restriction of φ to V_y is a continuous bijection onto its image with a continuous inverse.

Example 1. Take X to be the unit circle C in $\mathbb{C}$ with the usual topology, let Y be an open interval $-\infty \leq a < t < b \leq +\infty$, and define $\varphi : (a, b) \longrightarrow C$ by $\varphi(t) = e^{it}$. Then, (a, b) is spread over C by the mapping φ.

From this example, we see that if $\varphi : Y \to X$ is a spread of Y over X, then φ need not be surjective. Moreover, for those students who know what a covering space is, this example shows that a spread over X need not be a covering space, even if it is surjective. However, *a covering space is always a spread.*

Problem 41. If Y is a connected Hausdorff space with countable base spread over a complex manifold X, then there is a unique complex structure on Y with respect to which Y is a complex manifold and the projection is holomorphic.

Under these circumstances, we say that Y is a (complex) *manifold spread* over the (complex) manifold X. We also say that Y is a *spread manifold* over X. The quotient M/G of a manifold with respect to a properly discontinuous automorphism group G which acts freely is an example. That is, M is a manifold spread over M/G by the natural projection. In the next section, we give another example, the Riemann domain of a holomorphic function.

Chapter 13
Holomorphic Continuation

Abstract In one variable, the problem of holomorphic continuation leading to multi-valued functions gave rise to the concept of a Riemann surface, on which these ambiguous functions become well defined. A similar line of thought in several variables leads to the notion of a Riemann domain.

13.1 Direct Holomorphic Continuation and Domains of Holomorphy

The notion of holomorphic continuation is familiar from the study of functions of a single complex variable and was defined in the introduction for functions of several complex variables. We repeat the remarks made in the introduction. Let f_j be holomorphic in a domain Ω_j, $j = 1, 2$ and suppose if $f_1 = f_2$ in some component G of $\Omega_1 \cap \Omega_2$, then f_2 is said to be a direct holomorphic *continuation* of f_1 through G. In shorthand, we also say (f_2, Ω_2) is a *direct holomorphic continuation of* (f_1, Ω_1).

In one variable, holomorphic continuation is usually done from one disc to another using power series. We can do the same with polydiscs. Recall that if a power series converges in a polydisc, we have shown that the sum is holomorphic in that polydisc.

Problem 42. Let f_1 be the sum of a power series converging in a polydisc $\mathcal{D}_1$ centered at a_1 and let $a_2 \in \mathcal{D}_1$. Let f_2 be the sum of the Taylor series of f_1 about a_2. Then, f_2 converges in any polydisc centered at a_2 and contained in $\mathcal{D}_1$. If $\mathcal{D}_2$ is any polydisc centered at a_2 in which f_2 converges, then $(f_2, \mathcal{D}_2)$ is a direct holomorphic continuation of $(f_1, \mathcal{D}_1)$.

Problem 43. Let f be holomorphic in a domain U, let Ω be a domain which meets U, and let G be a component of the intersection. Show that, if there is a direct holomorphic continuation of f to Ω through G, then it is unique.

Let f be holomorphic in a domain Ω and let $p \in \partial\Omega$. We say that f has a direct holomorphic continuation to p if there is a holomorphic function f_p in a neighborhood U_p of p such that (f_p, U_p) is a direct holomorphic continuation of (f, Ω) through some component G of $\Omega \cap U_p$ with $p \in \partial G$.

P.M. Gauthier, *Lectures on Several Complex Variables*,
DOI 10.1007/978-3-319-11511-5_13

A domain Ω is a *domain of holomorphy* if it is the "natural" domain for some holomorphic function. That is, if there is a function f holomorphic in Ω which cannot be directly holomorphically continued to any boundary point of Ω. In particular, f cannot be directly holomorphically continued to any domain which contains Ω.

In the introduction, it was left as an exercise to show that in $\mathbb{C}$, each domain is a domain of holomorphy.

An important difference between complex analysis in one variable and in several variables is the existence of domains which are not domains of holomorphy in $\mathbb{C}^n$, $n > 1$.

One of the major problems in the history of several complex variables was to characterize domains of holomorphy. Let us say that a domain Ω is not holomorphically extendable at a boundary point p, if there is a neighborhood U_p of p and function f holomorphic in $\Omega \cap U_p$ which is not holomorphically extendable to the point p. Clearly, a domain of holomorphy has the property that, at each of its boundary points, it is not holomorphically extendable. E. E. Levi showed that the domain has a property called pseudoconvexity. The inverse problem, that is, the problem of showing that pseudoconvex domains are domains of holomorphy, was formulated by Levi in 1910 and is known as the Levi problem. This problem turned out to be extremely difficult. Oka was finally able to solve the problem in 1942 by introducing plurisubharmonic functions, which were also independently discovered by Lelong at about the same time. Because of research in this direction, pseudoconvexity became the single most important idea in the subject.

To describe domains of holomorphy in terms of pseudoconvexity, we shall first give an example of a domain which is not a domain of holomorphy and then use this example to help define a pseudoconvex domain. For simplicity, we shall restrict the discussion to $\mathbb{C}^2$. For $0 < \epsilon < 1$, we define the Hartogs figure

$$H_\epsilon = \{(z_1, z_2) \in \mathbb{D}^2 : |z_1| < \epsilon, \text{ or } 1 - \epsilon < |z_2| < 1\}.$$

In Theorem 15 we showed that H_ϵ is not a domain of holomorphy.

By a parametric bidisc in $\mathcal{D} \subset \mathbb{C}^2$, we mean a biholomorphic image of the unit bidisc $\mathbb{D}^2$. That is, there is a biholomorphic mapping $\Phi : \mathbb{D}^2 \to \mathcal{D}$. Let us say that $\Gamma \subset \mathbb{C}^2$ is a parametric Hartogs figure if there is such a parametric bidisc and a Hartogs figure $H_\epsilon \subset \mathbb{D}^2$ such that $\Phi(H_\epsilon) = \Gamma$.

We shall say that a domain $\Omega \subset \mathbb{C}^2$ is *pseudoconvex* if for every parametric bidisc $\mathcal{D} \subset \mathbb{C}^2$ and associated parametric Hartogs figure $\Gamma \subset \mathcal{D}$, if $\Gamma \subset \Omega$, then also $\mathcal{D} \subset \Omega$. There are many different definitions of pseudoconvexity and this one has frequently been called Hartogs pseudoconvexity for obvious reasons.

Theorem 55. *If $\Omega \subset \mathbb{C}^2$ is a domain of holomorphy, then Ω is Hartogs pseudoconvex.*

Proof. Suppose $f \in \mathcal{O}(\Omega)$. If Ω is not pseudoconvex, then there is a biholomorphism $\Phi : \mathbb{D}^2 \to \mathcal{D} \subset \mathbb{C}^2$ and a Hartogs figure $H \subset \mathbb{D}^2$ such that $\Phi(H) :\equiv \Gamma \subset \Omega$ but $\mathcal{D} \not\subset \Omega$. Since $\mathcal{D}$ is path connected, there is a point $p \in \partial\Omega \cap \mathcal{D}$.

Since $f \circ \Phi \in \mathcal{O}(H)$, it follows from Theorem 15 that $f \circ \Phi$ extends to a function $g \in \mathcal{O}(\mathbb{D}^2)$. On $\mathcal{D}$, we may define the holomorphic function $F = g \circ \Phi^{-1}$. For $z \in \Gamma$, there is a unique corresponding point $\zeta = \Phi^{-1}(z) \in H$. We have, for $z \in \Gamma$,

$$F(z) = F(\Phi(\zeta)) = (g \circ \Phi^{-1})(\Phi(\zeta)) = g(\zeta) = (f \circ \Phi)(\zeta) = f(z).$$

Since Γ is an open connected subset of $\mathcal{D} \cap \Omega$, it follows from the uniqueness principle that $F = f$ on the component of $\mathcal{D} \cap \Omega$ containing Γ. Thus, f extends holomorphically to the boundary point $p \in \partial\Omega$. Since f was an arbitrary function in $\mathcal{O}(\Omega)$, we have that every $f \in \mathcal{O}(\Omega)$ extends holomorphically to the boundary point p. We have shown that, if Ω is not Hartogs pseudoconvex, then it is not a domain of holomorphy, which concludes the proof. $\square$

We introduce another form of pseudoconvexity. Recall that by Theorem 45 a C^2-function r is convex in a convex domain if and only if $H_r \geq 0$. Thus, r is locally convex in a domain if and only if $H_r \geq 0$. It is thus reasonable to define a C^2-function $r : U \to \mathbb{R}$ to be locally *strictly* convex if $H_r > 0$. With this in mind, let us define a bounded domain $\Omega \subset \mathbb{R}^n$ to be *strictly convex* if there is a neighborhood U of $\partial\Omega$ and a strictly convex C^2-function $r : U \to \mathbb{R}$ such that $\Omega \cap U = \{x \in U : r(x) < 0\}$, $\partial\Omega = \{x \in U : r(x) = 0\}$, and $\nabla r(x) \neq 0$, for $x \in \partial\Omega$. By choosing a smaller U, we may assume that $\nabla r(x) \neq 0$, for all $x \in U$. To say that $H_r > 0$ in U means that, for every $x \in U$, and every nonzero $\xi \in \mathbb{R}^n$, we have $\sum_{j,k}(\partial^2 r/\partial x_j \partial x_k)(x)\xi_j\xi_k > 0$. In particular, if we choose ξ as the n-tuple having 1 in position j, and 0 elsewhere, we conclude that $(\partial^2 r/x_j^2)(x) > 0, j = 1, \cdots, n$. Thus, u is strictly subharmonic on U. Recall also that a C^2-function r is strictly plurisubharmonic in a domain of $\mathbb{C}^n$ if and only if it is locally strictly plurisubharmonic and that this is the case if and only if $L_r > 0$. In analogy with the definition of a strictly convex domain in $\mathbb{R}^n$, let us say that a bounded domain $\Omega \subset \mathbb{C}^n$ is *Levi strictly pseudoconvex* if there is a neighborhood U of $\partial\Omega$ and a function r strictly plurisubharmonic in U, such that $\Omega \cap U = \{z \in U : r(z) < 0\}$, $\partial\Omega = \{z \in U : r(z) = 0\}$ and $\nabla r(z) \neq 0$, for all $z \in U$.

Example. The unit ball $\mathbb{B} \subset \mathbb{C}^n$ is Levi strictly pseudoconvex. Indeed, for $r(z) = |z|^2 - 1$, we have $\mathbb{B} = \{z : r(z) < 1\}$. Moreover, $H_r = (\partial^2 r/\partial z_j \partial\bar{z}_k)$ is the identity matrix I, so

$$H_r(z)(\zeta) = \sum_{j,k=1}^{n} \frac{\partial^2 r}{\partial z_j \partial\bar{z}_k}(z)\zeta_j\bar{\zeta}_k = \sum_{i=1}^{n} \zeta_i\bar{\zeta}_i = |\zeta|^2 > 0, \quad \text{if } \zeta \neq 0.$$

Also, $\nabla r(z) = (2x_1, 2y_1, \cdots, 2x_n, 2y_n) \neq 0$, if $z \neq 0$. Thus, r is strictly plurisubharmonic and consequently the ball is Levi strictly pseudoconvex.

The notion of Levi strictly pseudoconvex domains was very important in solving the Levi problem. For a complete characterization of domains of holomorphy in terms of pseudoconvexity, we refer to the standard texts on several complex variables in the bibliography. Also, for a nice survey on pseudoconvexity and the Levi problem, see [15].

13.2 Indirect Holomorphic Continuation and Riemann Domains

Having considered *direct* holomorphic continuation, we now introduce *indirect* holomorphic continuation. A *holomorphic element* is a pair (f, U), where U is an open set in $\mathbb{C}^n$ and f is a function holomorphic in U. By *holomorphic continuation along a chain of elements*, we understand a sequence $(f_1, U_1), (f_2, U_2), \cdots, (f_m, U_m)$ of holomorphic elements, with given components G_j of successive intersections, such that (f_{j+1}, U_{j+1}) is a direct holomorphic continuation of (f_j, U_j) through G_j, for $j = 1, \cdots, m - 1$. We say that *there is a holomorphic continuation along a chain of elements from an element* (f, U) *to an element* (h, Ω) if there is a holomorphic continuation along a chain whose first element is (f, U) and whose last element is (h, Ω). We say that *a holomorphic element* (h, Ω) *is a holomorphic continuation of a holomorphic element* (f, U) if there is a holomorphic continuation along a chain of elements from the element (f, U) to the element (h, Ω). We say that (h, Ω) is an *indirect* holomorphic continuation of (f, U) if (h, Ω) is a holomorphic continuation of (f, U) but is not a direct holomorphic continuation of (f, U).

Problem 44. Give an example of an indirect holomorphic continuation.

Problem 45. Holomorphic continuation along a given chain of domains through a given sequence of components of the respective intersections is unique. That is, if $(f_1, U_1), \cdots, (f_m, U_m)$ and $(g_1, U_1), \cdots, (g_m, U_m)$ are holomorphic continuations along the same chain of domains $U_1, \cdots, U_m$ through the same components G_j of successive intersections with same initial functions $f_1 = g_1$, then the terminal functions are the same $f_m = g_m$.

Problem 46. Describe the Riemann surface of the logarithm.

We shall now introduce the notion of the *Riemann domain* of a holomorphic function of several complex variables. It is the higher dimensional analog of the Riemann surface of a function of a single complex variable. We shall see that the Riemann domain of a function is a special case of a spread manifold. All of this subsection could be simplified, if we were willing to assume some familiarity with sheaf theory (see [8]).

To construct the Riemann domain of a holomorphic element (f, U), we shall consider all holomorphic continuations along chains (f_j, U_j) through components G_j of successive intersections with initial element (f, U) equal to (f_1, U_1) and we shall "glue" successive domains along the sets G_j. In the case of one variable, this is the familiar construction of the Riemann surface of a holomorphic element. For functions of several variables, the procedure is the same. It turns out to be a spread

manifold constructed by this gluing process. Having sketched the construction of
the Riemann domain of a holomorphic element (f, U), we shall now explain the
process more carefully.

Let $f \in \mathcal{O}(\Omega)$, where Ω is a domain in $\mathbb{C}^n$. Let $\pi : M \to \mathbb{C}^n$ be a manifold
spread over $\mathbb{C}^n$. We shall say that M contains Ω if there is a domain $\tilde{\Omega}$ in M such
that π maps $\tilde{\Omega}$ biholomorphically onto Ω. By *the Riemann domain of f* we mean
the maximal complex manifold spread over $\mathbb{C}^n$ containing Ω to which f extends
holomorphically. Let us denote the Riemann domain of f by M_f. Thus, M_f is the
natural domain of f (over $\mathbb{C}^n$). From the definition of domain of holomorphy, we
have the following.

Theorem 56. *A domain $\Omega \subset \mathbb{C}^n$ is a domain of holomorphy if and only if there
exists $f \in \mathcal{O}(\Omega)$ such that $M_f = \Omega$.*

At this point, it would be good for the student to reconsider the earlier problem
of describing the Riemann surface of the logarithm. Fix a branch f of $\log z$ in some
domain Ω of $\mathbb{C}$. Describe the Riemann surface M_f.

Let us now return to the task of describing the Riemann domain M_f of a
holomorphic function $f \in \mathcal{O}(\Omega)$, where Ω is a domain in $\mathbb{C}^n$. Since M_f will
be associated to all possible holomorphic continuations of f, the end result will be
the same if we construct the Riemann domain of the restriction of f to some ball
B contained in Ω, for the direct holomorphic continuation of f from B to Ω is
unique. We thus assume from the outset that we are given an (holomorphic) element
(f, B). By this we mean that B is a ball in $\mathbb{C}^n$ and $f \in \mathcal{O}(B)$. We shall construct
the Riemann domain M_f of the element (f, B). Let

$$\mathcal{F} = \bigsqcup \{(f, B) : B \subset \mathbb{C}^n, \ f \in \mathcal{O}(B)\}$$

be the disjoint union of all elements (f, B) for all balls B in $\mathbb{C}^n$. We consider $\mathcal{F}$
as a topological space by putting the topology of the ball B on each (f, B). It may
help to think of f as a mere index on the ball B. Thus, the topological space $\mathcal{F}$ is
a disjoint union of balls (f, B). The ball (f, B) can be considered to be over the
ball B. We merely map the point (f, z) of (f, B) to the point $z \in B$. Thus, the
topological space $\mathcal{F}$ is spread over $\mathbb{C}^n$. Note that $\mathcal{F}$ is not connected. All of the balls
(f, B) are disjoint from each other and are in fact distinct components of $\mathcal{F}$.

Let us now introduce an equivalence relation on $\mathcal{F}$. Let $(f_\alpha, z_\alpha) \in (f_\alpha, B_\alpha)$ and
$(f_\beta, z_\beta) \in (f_\beta, B_\beta)$ be any two points in $\mathcal{F}$. We write $(f_\alpha, z_\alpha) \sim (f_\beta, z_\beta)$ if $z_\alpha = z_\beta$
and $f_\alpha = f_\beta$ on $B_\alpha \cap B_\beta$.

Problem 47. This is an equivalence relation on $\mathcal{F}$.

Let us denote the quotient space $\mathcal{O} = \mathcal{F}/\sim$. We describe the preceding
construction by saying that the space $\mathcal{O}$ is obtained by gluing two balls (f_α, B_α)
and (f_β, B_β) in $\mathcal{F}$ together along the intersection $B_\alpha \cap B_\beta$ if and only if $f_\alpha = f_\beta$
on this intersection. Let us denote an element of $\mathcal{O}$ by $[f, z]$, where $[f, z]$ is the
equivalence class of the element (f, z) in $\mathcal{F}$. An element $[f, z]$ is called *a germ of
a holomorphic function at z* and $\mathcal{O}$ is called the *space of germs of holomorphic*

functions. We claim that the space $\mathcal{O}$ is Hausdorff. Let $[f_\alpha, z_\alpha]$ and $[f_\beta, z_\beta]$ be distinct points of $\mathcal{O}$ representing, respectively, equivalence classes of points (f_α, z_α) and (f_β, z_β) in $\mathcal{F}$. Since $[f_\alpha, z_\alpha] \neq [f_\beta, z_\beta]$, either $z_\alpha \neq z_\beta$ or $z_\alpha = z_\beta$, but f_α and f_β are distinct holomorphic functions in some neighborhood B of $z_\alpha = z_\beta$.

In the first case, we may choose disjoint small balls U_α and U_β containing z_α and z_β and contained in the domains B_α and B_β of f_α and f_β, respectively. Since U_α and U_β are disjoint, the disjoint open sets (f_α, U_α) and (f_β, U_β) in $\mathcal{F}$ remain disjoint in $\mathcal{O}$. These yield disjoint neighborhoods of $[f_\alpha, z_\alpha]$ and $[f_\beta, z_\beta]$.

In the second case, no points of (f_α, B) and (f_β, B) are identified, for this would imply that $f_\alpha = f_\beta$ in B, which is not the case. Thus, (f_α, B) and (f_β, B) are disjoint neighborhoods of $[f_\alpha, z_\alpha]$ and $[f_\beta, z_\beta]$, respectively. We have shown that $\mathcal{O}$ is Hausdorff.

At last, we may define the Riemann domain M_f of an arbitrary holomorphic element (f, Ω), that is, of an arbitrary holomorphic function defined on a domain Ω in $\mathbb{C}^n$. Let B be a ball in Ω and define the Riemann domain M_f of f (more precisely, of (f, Ω)) to be the component of $\mathcal{O}$ containing the element (f, B).

It can be verified that the Riemann domain M_f of a holomorphic element (f, B) is indeed a manifold. Since M_f is spread over $\mathbb{C}^n$, we have only to check that M_f is Hausdorff and connected and has a countable base. First of all M_f is connected by definition, since it is a component of $\mathcal{O}$. To check that M_f is Hausdorff, it is sufficient to note that $\mathcal{O}$ is Hausdorff, since M_f is a subspace of $\mathcal{O}$, which we have shown to be Hausdorff.

To show that the Riemann domain M_f of a holomorphic element (f, B) is second countable is not so simple. We shall merely sketch the proof. It will be sufficient to construct a second countable subset X_f of M_f which is both open and closed. Since M_f is connected it will follow that $X_f = M_f$ so M_f is second countable.

We shall define the Riemann domain associated to holomorphic continuation of an element along chains as in one complex variable. Let

$$(f_1, B_1), \cdots, (f_m, B_m)$$

be a holomorphic continuation along a chain of balls $B_1, \cdots, B_m$. We construct a spread manifold associated to this holomorphic continuation by gluing two balls (f_j, B_j) and (f_k, B_k) along their intersection if and only if their intersection is nonempty and $f_j = f_k$ on this intersection. The resulting space is connected since in this process any two successive balls are glued together. This yields a complex manifold spread over $\mathbb{C}^n$. Using the same gluing rule, we may construct a complex manifold from any two holomorphic continuations along balls

$$(f_1, B_1), \cdots, (f_m, B_m) \quad \text{and} \quad (g_1, K_1), \cdots, (g_\ell, K_\ell)$$

having the same initial element (f_1, B_1). That is, the balls B_1 and K_1 are the same and the holomorphic functions f_1 and g_1 coincide thereon. We can do the same with any finite collection of holomorphic continuations having the same initial element (f, B). The result will always be connected and hence a complex manifold spread over $\mathbb{C}^n$.

Now let us perform such holomorphic continuations in a more systematic manner. Fix an initial holomorphic element (f, B). Let $M_{j,k}$ be the complex manifold spread $\mathbb{C}^n$ obtained by holomorphic continuation along chains of at most k balls of radius $1/j$ whose centers are obtained, starting from the center of B by taking (at most k) steps of length $1/j$ in the directions of the coordinate axes. Now we glue two such manifolds together along the intersection of two of their balls according to the usual rule. If we do this simultaneously to the whole family $M_{j,k}$, where $j, k = 1, 2, 3, \cdots$, the result is connected and yields a second countable Hausdorff space X_f spread over $\mathbb{C}^n$. Now, if M_f is the Riemann domain of a holomorphic element (f, B) and X_f is the manifold associated to holomorphic continuation of the element (f, B) constructed in the manner we have just described, it is not difficult to see that X_f is an open and closed subset of M_f. Since M_f is connected, $M_f = X_f$. Since X_f is by construction second countable, we have that M_f is second countable. Thus, the Riemann domain of a holomorphic element (f, B) is indeed a manifold.

Holomorphic continuation of a holomorphic element (f, B) usually leads to a multiple-valued "function." That is, if by continuation along a chain we return to a former point, the new function may differ from the former function at that point. The Riemann domain M_f of a holomorphic element (f, B) is constructed to remove this ambiguity. That is, there is a holomorphic function $\tilde{f}$ on M_f such that $\tilde{f} = f$ on the initial B. The function $\tilde{f}$ is defined on any (f_α, B_α) arising in the definition of M_f, by setting $f = f_\alpha$. The equivalence relation used in defining M_f is designed precisely to assure that $\tilde{f}$ is well defined, that is, $\tilde{f}$ is a (single-valued) function on M_f. We recapitulate by once again saying that the Riemann domain M_f of the holomorphic element (f, B) is the *natural domain of f*, that is, the maximal domain over $\mathbb{C}^n$ to which f extends holomorphically.

Chapter 14
The Tangent Space

Abstract The complex tangent space to a complex manifold allows us to define binary forms. Various binary forms are used to define Hermitian, symplectic, and almost complex manifolds.

The notion of complex tangent space is less intuitive than that of a real tangent space, just as complex numbers are less intuitive than real numbers. We shall begin by giving an intuitive interpretation of the real tangent space. Then, we shall formally define the real tangent space using derivations. The complex tangent space will be presented as a formal analog of the real tangent space. This is similar to introducing the complex numbers as formal expressions $a + ib$, where $a, b \in \mathbb{R}$. This approach to the tangent space via derivations is well presented in [11] and [18].

Here is a short definition which we shall explain afterwards. *The tangent space $T(X)$ of a real manifold X* of dimension n is the set of formal expressions in local coordinates

$$T(X) = \left\{ a_1 \frac{\partial}{\partial x_1} + \cdots + a_n \frac{\partial}{\partial x_n} : a_j \in C^1(X) \right\},$$

which is the space of smooth vector fields on X. A certain compatibility of these expressions is required with respect to change of local coordinates. We shall define the tangent space $T_p(X)$ of X at a point $p \in X$ and we shall set $T(X)$ equal to the disjoint union:

$$T(X) = \bigsqcup_{p \in X} T_p(X).$$

There remains to define $T_p(X)$.

Let X be a smooth (real) manifold. If U is an open subset of X, we denote by $\mathcal{E}(U)$ the set of smooth functions on U. If $p \in X$, let us say that f is a smooth function at p if $f \in \mathcal{E}(U)$ for some open neighborhood U of p. Two smooth functions f and g at p are said to be equivalent if $f = g$ in some neighborhood of p. This is an equivalence relation and the equivalence classes are called germs of

© Springer International Publishing Switzerland 2014
P.M. Gauthier, *Lectures on Several Complex Variables*,
DOI 10.1007/978-3-319-11511-5_14

smooth functions at p. For simplicity, we shall denote the germ of a smooth function f at p also by f. Denote by $\mathcal{E}_p$ the set of germs of smooth functions at p. The set $\mathcal{E}_p$ is an $\mathbb{R}$-algebra.

A *derivation* of the algebra $\mathcal{E}_p$ is a vector space homomorphism

$$D : \mathcal{E}_p \to \mathbb{R}$$

such that, if $f, g \in \mathcal{E}_p$, then

$$D(fg) = D(f) \cdot g(p) + f(p) \cdot D(g),$$

where $f(p)$ and $g(p)$ are the evaluations at p of the germs f and g.

The tangent space of X at p is the vector space of derivations of the algebra $\mathcal{E}_p$ and is denoted by $T_p(X)$.

Let

$$\varphi : U \to W$$

be a diffeomorphism of an open neighborhood U of p onto an open set $W \subset \mathbb{R}^n$ and set $\varphi^* f(x) = f \circ \varphi(x)$. Then, for every open $V \subset W$,

$$\varphi^* : \mathcal{E}(V) \to \mathcal{E}(\varphi^{-1}(V))$$

is an algebra isomorphism. Thus φ^* induces an algebra isomorphism on germs:

$$\varphi^* : \mathcal{E}_{\varphi(p)} \to \mathcal{E}_p,$$

and consequently induces an isomorphism on derivations:

$$\varphi_* T_p(X) \to T_{\varphi(p)}(\mathbb{R}^n).$$

Namely, for $D \in T_p(X)$, we define $\varphi_*(D) \in T_{\varphi(p)}(\mathbb{R}^n)$ as follows: we set

$$\varphi_*(D)f \equiv D(\varphi^* f), \quad \text{for} \quad f \in \mathcal{E}_{\varphi(p)}.$$

Problem 48. Fix $a \in \mathbb{R}^n$. Then,

(i) $\frac{\partial}{\partial x_1}, \cdots, \frac{\partial}{\partial x_n}$ are derivations of $\mathcal{E}_a(\mathbb{R}^n)$ and
(ii) form a basis of $T_a(\mathbb{R}^n)$.

Applying this to the point $a = \varphi(p)$, we see that the vector space $T_p(X)$ is of dimension n for each $p \in X$. The derivations given in the previous problem are the directional derivatives evaluated at the point $\varphi(p)$.

Let $f : M \to N$ be a smooth mapping between smooth manifolds. Then, there is a natural mapping:

$$f^* : \mathcal{E}_{f(p)} \to \mathcal{E}_p,$$

which in turn induces a natural mapping

$$df_p : T_p(M) \to T_{f(p)}(N)$$

given by $df_p(D_p) = D_p \circ f^*$. The mapping df_p is linear.

In local coordinates x for p and y for $q = f(p)$, consider the case $D_p = \frac{\partial}{\partial x_i}$. Let $g \in \mathcal{E}_q$:

$$\left(df_p(\frac{\partial}{\partial x_i})\right) g = (\frac{\partial}{\partial x_i} \circ f^*)g = \frac{\partial}{\partial x_i}(g \circ f) = \sum_{j=1}^{m} \frac{\partial g}{\partial y_j}\frac{\partial f_j}{\partial x_i} = \left(\sum_{j=1}^{m} \frac{\partial f_j}{\partial x_i}\frac{\partial}{\partial y_j}\right) g.$$

Thus

$$df_p(\frac{\partial}{\partial x_i}) = \sum_{j=1}^{m} \frac{\partial f_j}{\partial x_i}\frac{\partial}{\partial y_j}.$$

This maps a basis element of $T_p(M)$ to a basis element of $T_{f(p)}(N)$. Hence, in local coordinates, the linear transformation

$$df_p : T_p(M) \to T_{f(p)}(N)$$

is represented by the matrix

$$df_p = \begin{pmatrix} \frac{\partial f_1}{\partial x_1} & \cdots & \frac{\partial f_1}{\partial x_n} \\ \cdot & \cdots & \cdot \\ \cdot & \cdots & \cdot \\ \cdot & \cdots & \cdot \\ \frac{\partial f_m}{\partial x_1} & \cdots & \frac{\partial f_m}{\partial x_n} \end{pmatrix}.$$

The coefficients $\frac{\partial f_j}{\partial x_i}$ are smooth functions of the local coordinate x. The mapping df_p (sometimes denoted $f'(p)$) has the following names: the *derivative mapping* the *differential*, the *tangent mapping*, and the *Jacobian* of f at a. *The tangent mapping at p is the linear approximation of the smooth mapping f around p.*

Example. Consider a curve $\gamma : [0, 1] \to N$. Let M be the smooth manifold $(0, 1)$ and assume that $\gamma : (0, 1) \to N$ is smooth. The tangent space $T_t(0, 1)$ of the manifold $(0, 1)$ at the point t is of dimension 1 and has the tangent vector d/dt as basis. From the above considerations $d\gamma_t : T_t(0, 1) \to T_{\gamma(t)}N$ and, under this mapping, the image of the tangent vector d/dt at t is given in local coordinates $y = (y_1, \cdots, y_m)$ for N at $\gamma(t)$ by the formula

$$d\gamma_t(\frac{d}{dt}) = \gamma'(t) = \sum_{j=1}^{m} y_j'(t)\frac{\partial}{\partial y_j}.$$

Since the $\frac{\partial}{\partial y_j}$ are basis vectors in local coordinates for the tangent space $T_{\gamma(t)}N$, we may briefly write the previous formula in local coordinates for $T_{\gamma(t)}N$:

$$d\gamma = \gamma' = (y_1', \cdots, y_m'). \tag{14.1}$$

If g is a smooth function on N, then we may compare how the tangent vector (derivation) acts on g using a different set of coordinates $x = (x_1, \ldots, x_m)$ at $\gamma(t)$:

$$d\gamma_t\left(\frac{d}{dt}\right)(g) = \gamma'(t)(g) = \sum_{j=1}^{m} y_j' \frac{\partial g}{\partial y_j} = \sum_{j=1}^{m} y_j' \sum_{k=1}^{m} \frac{\partial g}{\partial x_k} \frac{\partial x_k}{\partial y_j} = \sum_{k=1}^{m} x_k' \frac{\partial g}{\partial x_k}. \tag{14.2}$$

We would like our definition of the tangent space $T_p(M)$ at a point p of a smooth manifold M to correspond to our intuitive notion of what it should be. The only situation in which we do have an intuitive notion is when we have an intuitive notion of M itself, that is, when M is a smooth submanifold of some Euclidean space. In this case, We think of $T_p(M)$ as the space of all vectors with base point p which are tangent to M at p. For our definition, it is preferable to think of these vectors as having the origin as base point, so that $T_p(M)$ is a vector subspace of the ambient Euclidean space. Tangent vectors generally are not contained in M, even if the base point p is. To obtain an intrinsic definition of $T_p(M)$, we note that there is a bijection between vectors a at the origin and derivatives $\sum a_j \frac{\partial}{\partial x_j}$ with respect to a. Moreover, this correspondence is preserved by smooth mappings and, in particular, by smooth change of charts. Namely, if f is a smooth mapping, σ is a smooth curve passing through p, and the vector a is tangent to σ at p, then since df_p is the linear approximation of f at p, the vector $df_p(a)$ is tangent to the curve $f \circ \sigma$ at $f(p)$.

Having discussed the tangent space to a smooth (real) manifold, we now introduce the (complex) tangent space to a complex manifold. Let p be a point of a complex manifold M and let $\mathcal{O}_p$ be the $\mathbb{C}$-algebra of germs of holomorphic functions at p. The *complex* (or *holomorphic*) *tangent space* $T_p(M)$ to M at p is the set of all derivations of the $\mathbb{C}$-algebra $\mathcal{O}_p$, hence the complex vector space homomorphisms $D : \mathcal{O}_p \to \mathbb{C}$ such that

$$D(fg) = f(p) \cdot D(g) + D(f) \cdot g(p).$$

For derivations $D_1, D_2, D \in T_p(M)$, and $\lambda \in \mathbb{C}$, we define the derivations $D_1 + D_2$ and λD in the obvious way and it is straightforward to verify that $T_p(M)$ is a complex vector space. In local coordinates, we note that $T_p(M) = T_z(\mathbb{C}^n)$ and that the partial derivatives $\{\frac{\partial}{\partial z_1}, \cdots, \frac{\partial}{\partial z_n}\}$ form a basis of $T_z(\mathbb{C}^n)$.

If M is a submanifold of some complex Euclidean space, then the complex tangent space to M at p is the largest complex vector subspace contained in the real tangent space to M at p. Having defined the (complex) tangent space $T_p(M)$

to a complex manifold at a point $p \in M$, we define the (complex) tangent space $T(M)$ of M:

$$T(M) = \bigsqcup_{p \in M} T_p(M).$$

There is a natural mapping $\pi : T(M) \to M$, defined by setting $\pi(v) = p$, for each $v \in T_p(M)$. We think of the tangent space $T(M)$ as lying "over" M and we think of the mapping $\pi : T(M) \to M$ as a "projection" of the tangent space $T(M)$ over M onto the space M "below."

Given a vector space V over a field F, a *multilinear* form of degree k is a mapping $\alpha : \prod_{j=1}^{k} V \to F$, which associates a scalar value $\alpha(v_1, \cdots, v_n)$ to each k-tuple $(v_1, \cdots, v_k)$ of vectors and which is linear in each variable, if the other $k - 1$ are fixed. For brevity, we shall call this a *k-form*. Real k-forms and complex k-forms are k-forms over the real and complex fields, respectively. Our main reason for presenting the tangent space TM to a manifold in these notes is to allow us to introduce several very important 2-forms, namely, Riemannian metrics, Hermitian metrics, fundamental forms, and symplectic forms.

14.1 Hermitian Manifolds

A *Hermitian* manifold (M, h) is a complex manifold M endowed with a Hermitian metric h, that is, a metric which is compatible with the complex structure. As usual, this is a complex analog of some real phenomenon, which is better known. Namely, a Hermitian manifold (M, h) is the complex analog of a *Riemannian* manifold (M, g), which is a "real" object.

We shall briefly recall the notion of a Riemannian manifold, which is, not surprisingly, a (real) manifold M endowed with a Riemannian metric g. But bear in mind that every complex manifold is also a real manifold (whose real dimension is twice its complex dimension). Thus, we can speak of Riemannian metrics on complex manifolds. The word *metric* has different meanings when speaking of Riemannian metrics or when speaking of metric spaces, but the two meanings are related and we shall explain the relation later. A *Riemannian metric* g on a smooth real manifold M assigns to each point $p \in M$ an inner product g_p on the tangent space $T_p(M)$ at p in such a way that g_p "varies smoothly" as p varies on M. In the tangent space $T_p(M)$, the notion of an inner product g_p on $T_p(M)$ certainly makes sense. The inner product g_p assigns to any two tangent vectors u and v at p a real number $g_p(u, v)$ in such a way that the mapping $g_p : T_p(M) \times T_p(M) \to \mathbb{R}$ has the properties required of an inner product. In local coordinates $x = (x_1, \cdots, x_n)$ at p, the g_p is an inner product on the tangent space $T_x \mathbb{R}^n$ to $\mathbb{R}^n$ at $x \in \mathbb{R}^n$. This tangent space is isomorphic to $\mathbb{R}^n$ and so in local coordinates, g_p is an inner product on $\mathbb{R}^n$ and hence is represented by a symmetric positive definite matrix $(g_{i,j})$. Thus, in local coordinates,

$$g_p(u, v) = \sum_{i,j} g_{i,j} u_i v_j$$

and we may define the length of a tangent vector $u \in T_p(M)$ to M at a point $p \in M$ as follows:

$$|u| = \sqrt{g_p(u, u)} = \left(\sum_{i,j} g_{i,j} u_i u_j \right)^{1/2}. \tag{14.3}$$

Now what does it mean for this inner product g_p to vary smoothly? First of all, let us say what it means for a vector to vary smoothly. A *vector field* u on M is a mapping $u : M \to T(M)$, which assigns to each point $p \in M$ a tangent vector $u(p) \in T_p(M)$. In local coordinates $x = x(p)$,

$$u(p) = \sum_{k=1}^{n} a_k(x) \frac{\partial}{\partial x_k},$$

and we say that u is a *smooth vector field* if the coefficients $a_k(x)$ are smooth functions. Now, we say that *an inner product g_p on $T_p(M)$ varies smoothly with p* if, whenever u and v are two smooth vector fields on M, the mapping

$$M \to \mathbb{R}$$

$$p \mapsto g_p(u(p), v(p))$$

is smooth. We can also describe the smoothness of g_p directly, without referring to smooth vector fields. To say that g_p varies smoothly is equivalent to saying that, in local coordinates, the elements $g_{i,j}(x)$ of this matrix vary smoothly with the local coordinate $x = x(p)$.

Putting this together, we say that *g is a Riemannian metric on a smooth manifold M* if g assigns to each $p \in M$ an inner product g_p on the tangent space $T_p(M)$ to M at p, and this inner product g_p varies smoothly with p. A *Riemannian manifold* (M, g) is a smooth real manifold M endowed with a Riemannian metric g.

Having defined the length of tangent vectors on Riemannian manifold (M, g), we shall use this to define the length of a smooth curve on M. Let us begin by reviewing the notion of length in Euclidean space $\mathbb{R}^n$. For two points $p, q \in \mathbb{R}^n$, we denote by $[p, q]$ the segment joining p and q. For a finite set of points $x_1, x_2, \cdots, x_m \in \mathbb{R}^n$, let us denote by $L[x_j : 1 \le j \le m]$ the polygonal line L obtained by joining successively the segments $[x_1, x_2], \cdots, [x_{m-1}, x_m]$. Then, the length $\ell(L)$ of the polygonal line L is given by $\ell(L) = \sum \ell[x_{j-1}, x_j] = \sum |x_j - x_{j-1}|$.

Given a curve $\gamma : [0, 1] \to \mathbb{R}^n$ in $\mathbb{R}^n$, let us say that L is a *polygonal approximation* of γ, if $L = L[x_j : 1 \leq j \leq m]$, where $x_j = \gamma(t_j)$, with $0 = t_1 \leq t_2 \leq \cdots \leq t_m = 1$. Set $|L| = \max |t_j - t_{j-1}|$. Let $\mathcal{L}(\gamma)$ denote the family of all polygonal approximations L of γ. It is natural to define the *length* $\ell(\gamma)$ *of the curve* γ as follows:

$$\ell(\gamma) = \limsup_{\epsilon \to 0} \{|L| : L \in \mathcal{L}(\gamma), |L| < \epsilon\}.$$

The length of γ may be infinite.

In case γ is a *smooth* curve in Euclidian space, we may calculate its *length* as the following integral:

$$\ell(\gamma) = \int_0^1 |\gamma'(t)| dt. \tag{14.4}$$

What is important here is that we are using the length of the tangent vector $\gamma'(t)$ at each point $\gamma(t)$ of the curve.

On a Riemannian manifold (M, g) we have all that we need to extend this notion of length of a curve. If $\gamma : [0, 1] \to M$ is a smooth curve in a Riemannian manifold (M, g), we define the length $\ell(\gamma)$ by the same formula (14.4), where, for each $t \in (0, 1)$, $\gamma'(t)$ is the associated tangent vector to M at the point $\gamma(t)$.

We need only say what we mean by the associated tangent vector $\gamma'(t)$. What we mean is the image by the mapping $d\gamma_t$ of the tangent vector d/dt, which is the canonical basis vector of the tangent space to $(0, 1)$ at the point t. In local coordinates on M, we may write the mapping γ as $\gamma(t) = (x_1(t), \cdots, x_n(t))$ and the tangent vector $\gamma'(t)$ is defined in these local coordinates as the vector $(x_1'(t), \cdots, x_n'(t))$. By formulas (14.1) and (14.3), then, we have as formula for the length of the curve γ

$$\ell(\gamma) = \int_0^1 |\gamma'(t)| dt = \int_0^1 \left(\sum_{i,j} g_{i,j} x_i' x_j' \right)^{1/2} dt.$$

Strictly speaking, since we have used local coordinates, this formula is valid if the curve is contained in a chart. However, we can express the curve as a finite concatenation $\gamma = \sum \gamma_\nu$, such that each γ_ν is contained in a chart. Then, we can calculate the length of each portion γ_ν using this formula and we can show that this is well defined, that is, does not depend on how we constructed the concatenation.

Now, having established a method of calculating the length $\ell(\gamma)$ of a (smooth) curve γ, we may define the distance $d(p, q)$ between two points p and q of the manifold as the infimum of lengths of smooth paths joining the two points. Thus, if $[p, q]$ is the class of smooth paths joining p and q, then

$$d(p, q) := \inf\{\ell(\gamma) : \gamma \in [p, q]\}, \quad p, q \in M.$$

Notice that we have implicitly used the fact that manifolds are path-connected. This follows from the fact that, by definition, they are connected and second countable. Let us verify that a Riemannian manifold (M, h) becomes a metric space with this distance. Certainly, for all $p, q \in M$, we have $d(p, q) = d(q, p)$ and $d(p, q) \geq 0$. It is easy to see that, for each point $p \in M$, we have $d(p, p) = 0$, since among the paths joining p and p, there is the constant path $\gamma(t) = p, 0 \leq t \leq 1$, for which the length is zero, and we have that $d(p, p) = 0$, for each $p \in M$. Now, suppose a, b, and c are points of M and, for $\alpha \in [a, b], \beta \in [b, c]$, denote by $\alpha + \beta$ the concatenation of α and β. Then,

$$d(a, c) = \inf\{\ell(\gamma) : \gamma \in [a, c]\} \leq \inf\{\ell(\alpha + \beta) : \alpha \in [a, b], \beta \in [b, c]\} =$$

$$\inf\{\ell(\alpha) + \ell(\beta) : \alpha \in [a, b], \beta \in [b, c]\} \leq$$

$$\inf\{\ell(\alpha) : \alpha \in [a, b]\} + \inf\{\ell(\beta) : \beta \in [b, c]\} = d(a, b) + d(b, c).$$

We have shown that the distance d induced by the metric g is indeed a distance function, so (M, d) is a metric space.

This completes our brief discussion of the notion of a Riemannian manifold (M, g) as a smooth (real) manifold endowed with a Riemannian metric g.

Since a complex manifold of dimension n is also a real manifold of (real) dimension $2n$, we may also consider a complex Riemannian manifold, that is, a complex manifold endowed with a Riemannian metric. However, for complex manifolds, there is a more interesting notion than a Riemannian metric, namely, a Hermitian metric to which we now turn.

Let V be a complex vector space. A *Hermitian inner product h* on V is a (complex) inner product on V. That is, h is a mapping $h : V \times V \to \mathbb{C}$, which satisfies the following, for $u, v, u_1, u_2, v_1, v_2 \in V$, and $\lambda \in \mathbb{C}$:

$$h(u_1 + u_2, v) = h(u_1, v) + h(u_2, v), \quad h(u, v_1 + v_2) = h(u, v_1) + h(u, v_2),$$

$$h(\lambda u, v) = \lambda h(u, v), \quad h(u, \lambda v) = \overline{\lambda} h(u, v), \quad h(v, u) = \overline{h(u, v)},$$

$$h(v, v) \geq 0 \quad \text{and} \quad h(v, v) = 0 \text{ iff } v = 0.$$

Notice that a Hermitian inner product is invariant under multiplication by i : $h(iu, iv) = h(u, v)$.

The standard inner product on $\mathbb{C}^n$ is $\langle z, w \rangle = \sum_{j=1}^{n} z_j \overline{w}_j$. On a Hermitian vector space (V, h), we may define a norm $|v| = \sqrt{h(v, v)}$ and hence a distance $|u - v|$ between vectors $u, v \in V$.

A *Hermitian matrix* is an $n \times n$ (complex) matrix $H = (h_{ij})$, such that $H = H^*$, where $H^* = (\overline{h}_{ji})$. *Every Hermitian inner product h on $\mathbb{C}^n$ can be represented by a Hermitian matrix H such that $h(z, w) = \sum h_{ij} z_i \overline{w}_j$. A square matrix is Hermitian if and only if all of its eigenvalues are real.*

A *Hermitian metric h* on a complex manifold M is an assignment, for each point $p \in M$, of an inner product h_p to the tangent space $T_p(M)$, which is Hermitian with respect to the complex structure on $T_p(M)$. We also assume that h_p varies smoothly as a function of p. A *Hermitian manifold* (M, h) is defined as a complex manifold M endowed with a Hermitian metric h.

Example Let C^n/L be the torus induced by a lattice L on $\mathbb{C}^n$. Since the projection $\mathbb{C}^n \to C^n/L$ is locally biholomorphic, it transfers the standard Hermitian metric h to $\mathbb{C}^n/L$. Thus, $(C^n/L, h)$ is a Hermitian manifold. However, although Euclidean space and the complex torus have the same Hermitian metric h, the induced distance functions are not the same. As a metric space, Euclidean space is unbounded while the torus is bounded. The torus with this standard Hermitian metric is called the *flat torus*.

If (M, h) is a Hermitian manifold, then we may associate to the Hermitian metric h a Riemannian metric g on the underlying real manifold, namely the real part of h defined as $g = (h + \bar{h})/2$. This Riemannian metric g is invariant under multiplication by $i : g(iu, iv) = g(u, v)$. Conversely, to any Riemannian metric g, we may associate a Hermitian metric. Firstly, we may define a Riemannian metric

$$\tilde{g}(u, v) = \frac{g(u, v) + g(iu, iv)}{2},$$

which is invariant under multiplication by i. Then, a direct verification shows that $h(u, v) = \tilde{g}(u, v) - i\tilde{g}(iu, v)$ yields Hermitian metric.

For example, consider the standard Hermitian metric on $\mathbb{C}^n$:

$$h(z, w) = \sum_{1 \leq j \leq n} z_j \bar{w}_j, \quad \text{with} \quad \bar{h}(z, w) = \sum_{1 \leq j \leq n} \bar{z}_j w_j.$$

Then, the underlying real vector space has coordinates $x_1, y_1, \cdots, x_n, y_n$, and

$$\frac{h(z, w) + \bar{h}(z, w)}{2} = \sum_{1 \leq j \leq n} (x_j u_j + y_j v_j) =$$

$$(x_1, y_1, \cdots, x_n, y_n) \cdot (u_1, v_1, \cdots, u_n, v_n),$$

which is the standard inner product on

$$\mathbb{R}^{2n} = \{(x_1, y_1, \cdots, x_n, y_n) : x_j, y_j \in \mathbb{R}\}.$$

Let M be a complex manifold. We shall say that *a Riemannian metric g on the underlying real manifold preserves the complex structure* if, for each point $p \in M$ and any two tangent vectors u, v at p, we have $g_p(iu, iv) = g_p(u, v)$. In the preceding paragraph we have seen that to each Hermitian metric on a complex manifold M, we can associate, in a natural way, a Riemannian metric on the underlying real manifold, which preserves the complex structure and, conversely, to each Riemannian metric on the underlying real manifold, which preserves the complex structure, we may associate, in a natural way, a Hermitian metric on M. In this sense, we may say that *a Hermitian manifold is a complex manifold whose underlying real manifold is assigned a Riemannian metric which preserves the complex structure.* To put it briefly to say that M is a Hermitian manifold is to say that we associate to M a complex bilinear form h, which is positive definite and conjugate symmetric, and a real bilinear form g which is positive definite and symmetric. Both forms preserve the complex structure: $h(iu, iv) = h(u, v)$ and $g(iu, iv) = g(u, v)$. We can recuperate each from the other $g = (h + \overline{h})/2$ and $h(u, v) = g(u, v) - ig(iu, v)$.

On a Hermitian manifold (M, h), we introduce a third bilinear form ω, called the *fundamental form*. It is defined as $\omega = \frac{i}{2}(h - \overline{h})$. The form ω is easily seen to be antisymmetric, $\omega(v, u) = -\omega(u, v)$, and it also preserves the complex structure, $\omega(iu, iv) = \omega(u, v)$. The three forms are related as follows: $h = g - i\omega$. Indeed,

$$g - i\omega = \frac{h + \overline{h}}{2} + \frac{h - \overline{h}}{2} = h.$$

A direct calculation verifies that ω and g can be obtained from each other as follows:

$$\omega(u, v) = g(iu, v)$$
$$g(u, v) = \omega(u, iv).$$

It follows from the second formula that $\omega(u, iv) > 0$ for any nonzero real tangent vectors u and v. Conversely, given an antisymmetric bilinear real form ω, which preserves complex structure and for which $\omega(u, iv) > 0$, when u and v are nonzero real tangent vectors, we obtain a Riemannian form g by setting $g(u, v) = \omega(u, iv)$. Indeed, g is symmetric, since

$$g(v, u) = \omega(v, iu) = -\omega(iu, v) = \omega(iu, -v) = \omega(iu, i^2 v) = \omega(u, iv) = g(u, v).$$

To summarize, we could say that a Hermitian structure on a complex manifold M is any one of the following three:

1. A Hermitian metric h
2. A Riemannian metric g which preserves the complex structure
3. An antisymmetric form ω which preserves the complex structure and for which $\omega(u, iu) > 0$ for every nonzero real tangent vector

14.2 Symplectic Manifolds

Symplectic manifolds play an important role in classical mechanics. Since it turns out that the real dimension of a symplectic manifold is necessarily even, they are also of importance in the study of complex manifolds.

A *symplectic manifold* (M, ω) is a smooth real manifold M equipped with a *symplectic form* ω, which by definition is a nondegenerate closed differential 2-form. A complex manifold is symplectic if the underlying real manifold is equipped with a symplectic form.

We suppose the student is familiar with *differential forms*. Recall that a differential 2-form on a real manifold of dimension n can be written in local coordinates as

$$\sum_{1 \le i < j \le n} a_{i,j} \, dx_i \wedge dx_j,$$

where the coefficients satisfy compatibility conditions for changing coordinates. A differential form ω is said to be *closed* if $d\omega = 0$, where d is the exterior derivative. We shall say momentarily what it means for a differential 2-form to be nondegenerate.

Given a smooth mapping $f : M \to N$ from one smooth real manifold to another, we have associated the differential mapping $df : T(M) \to T(N)$, between the associated tangent spaces, which for each $p \in M$ is a linear mapping $(df)(p) : T_p(M) \to T_{f(p)}(N)$ from the tangent space to M at the point p to the tangent space to N at the point $f(p)$. In particular, for a coordinate mapping $x = x(p)$, of a manifold of dimension n, we have the differentials dx_j, $j = 1, \cdots, n$. A differential form of degree k is given in local coordinates as

$$\sum_{j_1 < \cdots < j_k} a_{j_1, \cdots, j_k} dx_{j_1} \wedge \cdots \wedge dx_{j_k},$$

where the coefficients satisfy compatibility conditions for changing coordinates. The forms $dx_{j_1} \wedge \cdots \wedge dx_{j_k}$, $j_1 < \cdots < j_k$ form a basis for the space of differential k-forms. The *raison d'être* for differential k-forms is to give a meaning to integration over submanifolds of dimension k. Standard integration formulas force the wedge product to be alternating: $dx_i \wedge dx_j = -dx_j \wedge dx_i$. A differential 2-form is thus the same as an alternating bilinear form. A bilinear form α on a vector space V is said to be *nondegenerate* if, for each nonzero vector $u \in V$, there is a nonzero vector $v \in V$, such that $\alpha(u, v) \neq 0$ and, for each nonzero vector $v \in V$, there is a nonzero vector $u \in V$, such that $\alpha(u, v) \neq 0$. A differential 2-form ω on M is said to be *nondegenerate* if, for each $a \in M$, the form ω_a seen as a bilinear form on $T_a(M)$ is nondegenerate.

Example. There is a standard linear model of a symplectic manifold, namely $M = \mathbb{R}^{2n}$ equipped with coordinates

$$x_1, \cdots, x_n, y_1, \cdots, y_n$$

and the differential form

$$\omega = dx_1 \wedge dy_1 + \cdots + dx_n \wedge dy_n,$$

or equivalently, it is more suggestive physically to think of the classical coordinates for position and momentum: $q_1, \cdots, q_n, p_1, \cdots, p_n$ in which case

$$\omega = dq_1 \wedge dp_1 + \cdots + dq_n \wedge dp_n.$$

To show that the 2-form ω is nondegenerate, we must show that if u and v are tangent vectors and $\omega(u, v) = 0$, for each v, then $u = 0$. Suppose then that $\omega(u, v) = 0$, for each tangent vector v. Then, in particular, it is 0, for each vector in the basis

$$\frac{\partial}{\partial x_1}, \cdots, \frac{\partial}{\partial x_n}, \frac{\partial}{\partial y_1}, \cdots, \frac{\partial}{\partial y_1}$$

for tangent vectors. Thus, for $j = 1, \cdots, n$,

$$0 = \omega(u, \frac{\partial}{\partial x_j}) = \sum_{k=1}^{n}(dx_k \wedge dy_k)(u, \frac{\partial}{\partial x_j}) =$$

$$\sum_{k=1}^{n}\left[dx_k(u)dy_k(\frac{\partial}{\partial x_j}) - dx_k(\frac{\partial}{\partial x_j})dy_k(u) \right] = -dy_j(u),$$

and similarly,

$$0 = \omega(u, \frac{\partial}{\partial y_j}) = dx_j(u).$$

Now, we may write

$$u = \sum_{j=1}^{n}\left(a_j \frac{\partial}{\partial x_j} + b_j \frac{\partial}{\partial y_j} \right),$$

where $dx_j(u) = a_j$ and $dy_j(u) = b_j$. But since $dx_j(u) = dy_j(u) = 0$, it follows that $u = 0$, so ω is indeed nondegenerate. We have verified that ω is a symplectic form. It is called the *standard symplectic form* on $\mathbb{R}^{2n}$. This example is extremely important, because the following theorem of Darboux tells us that *every* symplectic form can be expressed as the standard symplectic form locally.

Theorem 57 (Darboux). *If ω is a symplectic form on a smooth real manifold M, and m is an arbitrary point of M, then there are local coordinates $x_1, \cdots, x_n, y_1, \cdots, y_n$, in a neighborhood of m, in terms of which ω is the standard symplectic form*

$$\omega = dx_1 \wedge dy_1 + \cdots + dx_n \wedge dy_n.$$

We shall not prove this theorem.

Recall that the fundamental form ω on a Hermitian manifold is antisymmetric (alternating) and nondegenerate and hence the fundamental form can be considered a nondegenerate differential 2-form. A fundamental form is thus a symplectic form if it is closed.

14.3 Almost Complex Manifolds

Almost complex manifolds are an attempt to endow a real manifold with a complex structure. If M is a complex manifold, at each point $p \in M$, the tangent space $T_p M$ to M at p has a natural structure as a complex vector space and multiplication by i in $T_p M$ is an automorphism of $T_p M$ such that $i^2 = -I$, where $I : T_p M \to T_p M$ is the identity mapping. In imitation of this, we define an *almost complex structure* J on a smooth real manifold M as an assignment, to every $p \in M$, of an automorphism $J_p : T_p M \to T_p M$, such that $J_p^2 = -I$. Moreover, we naturally ask that J_p vary continuously as p varies on M.

As usual, let us start with the linear situation, that is, vector spaces. Given a real (finite-dimensional) vector space V, we attempt to endow V with a complex structure, that is, the structure of a complex vector space. Let $I : V \to V$ be the identity transformation and let $J : V \to V$ be a linear transformation such that $J^2 = -I$. We call J an almost complex structure on V. Let us use J to construct a complex structure on V. We define complex scalar multiplication as follows. For $\lambda = \alpha + i\beta \in \mathbb{C}$, and v a vector in V, we set $\lambda v = \alpha v + \beta J v$. It is easy to check that complex scalar multiplication so defined makes V into a complex vector space. That is, J endows V with a complex structure. We may denote this complex vector space by (V, J).

Conversely, if V is a complex vector space, then V is also a real vector space and, if we define $J : V \to V$, by $Jv = iv$, then J is an almost complex structure on the underlying real vector space and the complex structure induced by J is the original complex structure on V.

If V admits an almost complex structure J, then we shall show that the (real) dimension n of V is even. Consider the n-dimensional polynomial $p(t) = \det[J - tI]$. If n is odd, then $p(t)$ has a real zero t_0. That is, $\det[J - t_0 I] = 0$. Hence $\ker(J - t_0 I) \neq \{0\}$. There is a nonzero vector $v \in \ker(J - t_0 I)$. Thus, $Jv = t_0 v$. Applying J to this equation, we have that $-v = t_0 Jv = t_0^2 v$. Consequently $t_0^2 = -1$. But t_0 belongs to $\mathbb{R}$ and it is impossible to have $t_0^2 = -1$. This gives a contradiction and thus n should be even. We have shown that if a real vector space admits an almost complex structure, then the dimension of V is even.

Let us show the converse, namely, that every finite-dimensional real vector space of even dimension admits an almost complex structure. The property of having an almost complex structure is invariant under isomorphisms. Indeed, suppose V and W are isomorphic real vector spaces and $T : V \to W$ is an isomorphism. If J is an almost complex structure for W, then it is easy to check that $T^{-1}JT$ is an almost complex structure for V. Thus, to show that all real even dimensional vector spaces admit an almost complex structure, it is sufficient to consider $\mathbb{R}^{2n}$, since every $2n$-dimensional real vector space is isomorphic to $\mathbb{R}^{2n}$.

Let $v_1, \cdots, v_{2n}$ be the standard coordinates for $v \in \mathbb{R}^{2n}$. Set

$$J_1 v = (-v_{n+1}, \cdots, -v_{n+n}, v_1, \cdots, v_n).$$

Then,

$$J_1^2 v = (-v_1, \cdots, -v_n, -v_{n+1}, \cdots, -v_{2n}) = -v,$$

so J_1 is an almost complex structure on $\mathbb{R}^{2n}$. Denote by $\mathbb{R}_1^{2n} = (\mathbb{R}^{2n}, J_1)$ the induced complex vector space. Let $e_1, \cdots, e_{2n}$ be the standard basis of $\mathbb{R}^{2n}$. We define vectors $e_j^1, j = 1, \cdots, n$ in $\mathbb{R}_1^{2n}$, by setting $e_j^1 = e_j, j = 1, \cdots, n$. Noting that, for $j = 1, \cdots, n$, we have $J_1 e_j = e_{j+n}$ and $J_1 e_{j+n} = -e_j$, suppose $\lambda_j = \alpha + i\beta_j \in \mathbb{C}$ and

$$0 = \sum_{j=1}^{n} \lambda_j e_j^1 = \sum_{j=1}^{n} (\alpha_j + i\beta_j) e_j = \sum_{j=1}^{n} (\alpha_j e_j - \beta_j J_1 e_j) = \sum_{j=1}^{n} (\alpha_j e_j - \beta_j e_{j+n}).$$

Since $e_1, \cdots, e_{2n}$ are $\mathbb{R}$-linearly independent, $\alpha_j = \beta_j = 0$ and hence $\lambda_j = 0, j = 1, \cdots, n$. Thus, the e_j^1 are $\mathbb{C}$-linearly independent and form a basis for the complex vector space $\mathbb{R}_1^{2n}$.

Now, consider

$$J_2 v = (-v_2, v_1, \cdots, -v_{2n}, v_{2n-1}).$$

Then,

$$J_2^2 v = (-v_1, -v_2, \cdots, -v_{2n-1}, -v_{2n}) = -v,$$

so J_2 is also an almost complex structure on $\mathbb{R}^{2n}$. Set $e_j^2 = e_{2j-1}, j = 1, \cdots, n$. Noting that $J_2 e_{2j-1} = e_{2j}, J_2 e_{2j} = -e_{2j-1}$, we can show, as in the previous paragraph, that the $e_j^2, j = 1, \cdots, n$, form a basis for the complex vector space $\mathbb{R}_2^{2n}$ induced by the almost complex structure J_2 on $\mathbb{R}^{2n}$.

To summarize, a finite-dimensional real vector space admits an almost complex structure J if and only if it is of even dimension and, in this case, the almost complex structure J is not unique. The two complex structures J_1 and J_2 which we have given on $\mathbb{R}^{2n}$ are *different* in the most obvious sense that $J_1 \neq J_2$. Let us say that two almost complex structures J_1 and J_2 induce the same complex structure on V if the identity mapping $(V, J_1) \to (V, J_2)$ is holomorphic. For the example of J_1 and J_2 we have given on $\mathbb{R}^{2n}$, they do *not* induce the same complex structure. To see this, we may define complex coordinates on $(\mathbb{R}^{2n}, J_1)$, by setting $z_j = v_j + iv_{n+j}$, $j = 1, \cdots, n$, and complex coordinates on $(\mathbb{R}^{2n}, J_2)$, by setting $w_j = v_{2j+1} + iv_{2j}$, $j = 1, \cdots, n$. The identity mapping on $\mathbb{R}^{2n}$ using these complex coordinates is

$$w_j = \frac{z_{2j-1} + \bar{z}_{2j-1} + z_{2j} - \bar{z}_{2j}}{2}, \quad j = 1, \cdots, n$$

which is clearly not holomorphic.

Although two almost complex structures J_1 and J_2 on a real vector space V may induce two *different* complex structures on V, in the sense that the identity mapping will not be holomorphic mapping, these two complex structures will be *equivalent* in the sense that there exists a biholomorphic mapping $(V, J_1) \rightarrow (V, J_2)$, not the identity in this case. To see that there exists a biholomorphic mapping, we note that any two vector spaces of the same dimension over the same field are isomorphic. Thus there exists a vector space isomorphism between the complex vector spaces (V, J_1) and (V, J_2). Since this isomorphism and its inverse are linear transformations, they are holomorphic.

We have just noted that, if V is a real *linear* space, all complex structures on V are equivalent. Such is not the case in the nonlinear situation, that is, on a manifold M. For example, on the disc in $\mathbb{R}^2$ (which is topologically equivalent to $\mathbb{R}^2$ itself), we may give the complex structure of the disc $\mathbb{D}$ in $\mathbb{C}$ or that of $\mathbb{C}$ itself. By Liouville's theorem, these two complex structures are not equivalent; there is no (nonconstant) holomorphic function $\mathbb{C} \rightarrow \mathbb{D}$.

Given a real submanifold X of an almost complex manifold (M, J) one can attempt to determine how much almost complex structure there is in X. We could define the holomorphic tangent space HT_pX at a point $p \in X$ to be the maximal subspace of the complex tangent space $\mathbb{C}T_pM$ to M at p which is contained in the real tangent space T_pX to X at p. Equivalently, $HT_pX = T_pX \cap JT_pX$. The dimension of HT_pX is called the Cauchy–Riemann dimension of X at the point p. It tells us how much almost complex structure X has at the point p. The real manifold X is called a Cauchy–Riemann manifold if its Cauchy–Riemann dimension is the same at all points of X. On Cauchy–Riemann manifolds, one can define Cauchy–Riemann functions, which are functions which try their best to be holomorphic. We shall not explain how this is done, for this so-called Cauchy–Riemann theory goes beyond the scope of these lecture notes. However, the interested student could consult the excellent survey [14].

Chapter 15
Meromorphic Functions and Subvarieties

Abstract This final chapter introduces two difficult subjects, which are unavoidable. We must study meromorphic functions if we are to deal with such simple "functions" as z/w. Moreover, we must study varieties, since the set of zeros of a holomorphic function is a variety. These two subjects are closely related, for the difficulty in studying a meromorphic "function" f/g arises from the zeros of the denominator.

15.1 Meromorphic Functions

It is not obvious how to define a meromorphic function on a complex manifold M. Let us first consider the most important meromorphic functions, namely rational functions. The set of rational functions on $\mathbb{C}^n$ is the quotient field $\mathbb{C}(z_1, \cdots, z_n)$ of the ring of polynomials $\mathbb{C}[z_1, \cdots, z_n]$ in n-variables. The quotient field is a purely algebraic concept, analogous to the construction of rational numbers from integers. We consider the family of pairs $f = (p, q)$, where $p, q \in \mathbb{C}[z_1, \cdots, z_n]$ and $q \neq 0$. Two such pairs (p_1, q_1) and (p_2, q_2) are said to be equivalent if $p_1 q_2 = p_2 q_1$. This is an equivalence relation and we define a rational "function" to be an equivalence class $[f] = [(p, q)]$ of such pairs. We shall sometimes simply use the notation f for the equivalence class $[f]$.

For $n = 1$, rational functions are indeed functions. We may assign a well-defined value $[f](z)$ to each z so that $[f]$ becomes a holomorphic mapping from the Riemann sphere to the Riemann sphere. For $n > 1$, it is not always possible to reasonably assign a value $[f](z)$ to a class $[f]$ and a point z. To see this, let us take $n = 2$ and consider $f = (p, q)$, where $p(z_1, z_2) = z_2$ and $q(z_1, z_2) = z_1$. We would like to assign the value z_2/z_1; however, it is not possible to do this in a continuous manner at the origin. Indeed, along an arbitrary complex line $z_2 = \lambda z_1, z_1 \neq 0, \lambda \in \mathbb{C}$, we have $f(z_1, z_2) = \lambda$ and so f assumes all values in every neighborhood of 0. We say that the origin is an indeterminate point for the rational function z_2/z_1.

We shall give two definitions for meromorphic functions, but this should not be too disturbing, because the class of meromorphic functions in the second sense will include all meromorphic functions in the first sense and add some more meromorphic functions that were "overlooked" at first glance.

Meromorphic definition 1. In imitation of the definition of rational functions, our first definition of meromorphic "functions" on a complex manifold M is to define $\mathcal{M}_1(M)$ as the quotient field of the ring $\mathcal{O}(M)$ of holomorphic functions on M. Since rational functions are instances of meromorphic functions, we see that a meromorphic "function" is not quite a function on M. But at least it will turn out to be a function on "most of" M.

Now we would like a meromorphic "function" on M to indeed be a function, that is, a rule which assigns to each point of the manifold M a complex value or infinity. Let $f = (g, h)$ be a pair of holomorphic functions on M, with $h \neq 0$. We shall occasionally denote the equivalence class $[f]$ merely by f or by the formal expression $f = g/h$. In a local coordinate z for $p \in M$, we shall also simply write $f(z)$ for $f(z(p))$. In an arbitrary chart U, g and h are holomorphic and by the uniqueness property of holomorphic functions $h|U \neq 0$, since by definition h is not the zero function on M. If $z \in U$ is a point where $h(z) \neq 0$, let us define $f(z)$ as the value $g(z)/h(z)$. Let us show that $f(z)$ is well defined. Suppose f has two representations $f = g_1/h_1 = g_2/h_2$ such that $h_1(z)$ and $h_2(z)$ are both different from 0. Then $h_1(z)h_2(z) \neq 0$. Since $g_1(z)h_2(z) = g_2(z)h_1(z)$, it follows that $g_1(z)/h_1(z) = g_2(z)/h_2(z)$. Thus, $f(z)$ is indeed well defined. Denote by $Z(h)$ the zero set of a holomorphic function $h \in \mathcal{O}(M)$. Then, f is well defined as a function on the open dense set $M \setminus Z(h)$. In fact, f is well defined on the union D_f of such sets

$$ D_f = \bigcup \{ M \setminus Z(h) : f = g/h, h \neq 0 \}. $$

At each point $p \in D_f$, the function f takes a well-defined complex value $f(p)$. Since D_f is open, if $p \in D_f$ and f has the representation g/h, with $h(p) \neq 0$, then the same is true for nearby points q. Thus $f(q) = g(q)/h(q)$, for q in a neighborhood of p and h has no zeros in this neighborhood of p. Hence, f is holomorphic in a neighborhood of p. We have shown that a meromorphic function $f \in \mathcal{M}_1(M)$ is holomorphic on the dense open subset $D_f \subset M$.

Now suppose that $p \in M$ and f has a representation g/h where $g(p) \neq 0$. We set $f(p) = g(p)/h(p)$, and we can show as before that, if f has the representations g_1/h_1 and g_2/h_2 with $g_1(p)$ and $g_2(p)$ different from 0, then $g_1(p)/h_1(p) = g_2(p)/h_2(p)$. If $h(p)$ is also different from 0, then we have already previously defined $f(p)$ and this "new" definition is consistent with the former. If this new definition adds any points to the domain of definition of f, they are points such that $g(p) \neq 0$, and $h(p) = 0$, for some representation g/h of f. We notice that $f(p) = \infty$ at such points and we call these points *poles*. Let us denote by P_f the set of poles of f. Thus, f is well defined at all points $p \in M$, for which f has a representation g/h, where either $g(p)$ or $h(p)$ is not zero and the set of such points is the union of two disjoint sets D_f and P_f. The function f is now seen to be holomorphic on the open dense set D_f and has poles at all points of the closed nowhere dense set P_f. We denote the remaining points of M by I_f and call them *indeterminate points* of f. Thus M is the disjoint union $M = D_f \cup P_f \cup I_f$.

If U and V are connected open subsets of M with $U \subset V$, then there is a natural restriction mapping $\mathcal{M}_1(V) \to \mathcal{M}_1(U)$, given by $(g, h) \mapsto (g|U, h|U)$, where $g, h \in \mathcal{O}(V)$ and $g \neq 0$. It is easy to see that this restriction mapping is well defined. That is,

$$(g_1, h_1) \sim (g_2, h_2) \Rightarrow (g_1|U, h_1|U) \sim (g_2|U, h_2|U).$$

From the uniqueness property of holomorphic functions, this restriction mapping is also injective. That is,

$$(g_1|U)(h_2|U) = (g_2|U)(h_1|U) \Rightarrow g_1 h_2 = g_2 h_1.$$

Thus, $\mathcal{M}_1(V)$ can be considered as a subfield of $\mathcal{M}_1(U)$. Since the assignment of values $f(p)$ to a meromorphic function is purely in terms of local behavior, it is preserved under this restriction mapping. Of course, the restriction mapping is in general not surjective. That is, there are usually more meromorphic functions on U than on V. We invite the student to consider the Hartogs phenomenon in this context.

A function f defined on a domain $\Omega \subset \mathbb{C}$ is holomorphic if and only if it is locally holomorphic. That is, if each point of Ω has an open neighborhood in which f is holomorphic, then f is holomorphic on Ω. The same is true if we replace Ω by a complex manifold. In one variable, a similar situation prevails for meromorphic functions. A function f is meromorphic on a complex manifold of dimension one (Riemann surface) if and only if it is meromorphic in a neighborhood of each point of the manifold. This property of meromorphic functions does not extend to higher dimensions, if we restrict the notion of meromorphic functions to that of the class $\mathcal{M}_1(M)$. The problem can be formulated as follows. Suppose, we are given a cover $U_\alpha, \alpha \in A$, of a complex manifold M by open connected sets. Suppose for each α we have a pair $f_\alpha = (g_\alpha, h_\alpha)$ of holomorphic functions on U_α, with $h_\alpha \neq 0$, and these pairs are compatible in the sense that $g_\alpha h_\beta = g_\beta h_\alpha$ on $U_\alpha \cap U_\beta$, whenever the latter is nonempty. Does it follow that there are (global) holomorphic functions g and h on M, with $h \neq 0$ such that $g h_\alpha = g_\alpha h$ on U_α, for each α? In shorthand notation, if we are given meromorphic functions $f_\alpha \in \mathcal{M}_1(U_\alpha)$, with $f_\alpha = f_\beta$ on $U_\alpha \cap U_\beta$, does there exist a (global) meromorphic function $f \in \mathcal{M}_1(M)$, with $f|U_\alpha \equiv f_\alpha$, for each α? The answer is in general no and is part of the famous *Cousin problems,* which are beyond the scope of the present lecture notes. We refer the student to the books in the bibliography.

The preceding dilemma can be finessed by giving a more general definition of meromorphic functions than $\mathcal{M}_1(M)$ which we now do. If p is a point of a complex manifold M, the ring $\mathcal{O}_p$ of germs of holomorphic functions at p is an integral domain and so we may form the quotient field, which we denote by $\mathcal{M}_p$. This field is called *the field of germs of meromorphic functions at p.* Thus, a meromorphic germ f_p at p can be represented as a quotient $f_p = g_p/h_p$, where $g_p, h_p \in \mathcal{O}_p$, and $h_p \neq 0$.

Meromorphic definition 2. Let us define a meromorphic function f on M as a mapping f which assigns to each $p \in M$ a meromorphic germ f_p at p. We impose the following compatibility between these germs. For every $p \in M$, there is a connected neighborhood U of p and holomorphic functions $g, h \in \mathcal{O}(U)$ with $h \neq 0$, such that $f_q = g_q/h_q$ for all $q \in U$. It turns out that we can (and shall) assume the following coherence property: for each $q \in U$, g_q and h_q are relatively prime. Let us denote by $\mathcal{M}(M)$ the class of meromorphic functions in this sense. It can be shown that $\mathcal{M}_1(M)$ *is a subclass of* $\mathcal{M}(M)$. *Whether or not the reverse inclusion holds on a particular manifold M is a Cousin problem.*

Now we would like a meromorphic function to indeed be a function, that is, taking complex values. Let $p \in M$ and U, g, and h be as in the above definition. We set

$$f(p) = \begin{cases} g(p)/h(p) & \text{if } h(p) \neq 0, \\ \infty & \text{if } h(p) = 0, g(p) \neq 0, \\ 0/0 & \text{if } h(p) = g(p) = 0. \end{cases}$$

The point p is called a *point of holomorphy* in the first case, a pole in the second case, and a *point of indetermination* in the third case. This trichotomy is well defined, that is, does not depend on the choice of U, g, and h. Moreover, the value $f(p)$ is also well defined in the first two cases. The first two cases form an open dense set G of M. In this sense, a meromorphic function is a well-defined function on most of M. In a neighborhood of each point of G, either f or $1/f$ is holomorphic. Note that f is being considered as a mapping in two ways. First of all, f was originally defined as a mapping which assigns to each $p \in M$ a germ f_p, with a compatibility condition between germs. Now we are also considering f as a mapping which assigns to each $p \in G$ a finite or infinite value $f(p)$, which may be considered as the value of the germ f_p at p. If p is a point of indetermination, then, for each complex value ζ, there are points of holomorphy q arbitrarily close to p such that $f(q) = \zeta$. For proofs of these claims, see [4].

That we have given two definitions for meromorphic functions should not be too disturbing. We have merely enlarged the class of meromorphic functions, since $\mathcal{M}_1(X) \subset \mathcal{M}(X)$. Moreover, the two definitions coincide for complex manifolds of dimension 1 (Riemann surfaces).

15.2   Subvarieties

Varieties are an important generalization of manifolds. We introduce them at this point, because the set of points of indetermination of a meromorphic function is an example of a variety and an understanding of varieties (with some outside reading) would help the student to better understand the behavior of meromorphic functions at points of indetermination, if he or she so chooses.

Zero sets of holomorphic functions in several variables are not always manifolds, but they are very close to being manifolds. Earlier, we gave the example of the holomorphic function of two complex variables $f(z, w) = zw$. The zero set is the union of the z-axis and the w-axis, which are both complex submanifolds. Thus, it is a manifold in the neighborhood of every one of its points except the origin. This is an example of an analytic subvariety.

Definition 58. A subset V of a complex manifold M is an analytic subvariety of M, if for every point $p \in M$, there is an open neighborhood U of p in M and a family $\mathcal{F}$ of functions holomorphic on U, such that

$$V \cap U = \{q \in U : f(q) = 0, \forall f \in \mathcal{F}\}.$$

In particular, *the zero set of a holomorphic function is an analytic subvariety*. It follows from the definition that subvarieties of M are closed subsets of M. Indeed, suppose $\{q_j\}$ is a sequence in V converging to a point $p \in M$. Let U and $\mathcal{F}$ be associated to p as in the definition. We may assume that the sequence $\{q_j\}$ lies in U. We have, for each $f \in \mathcal{F}$,

$$f(p) = \lim_{j \to \infty} f(q_j).$$

Hence $f(p) = 0$ for each $f \in \mathcal{F}$. Thus, $p \in V \cap U \subset V$ and so V is closed in M.

Theorem 59. *Submanifolds are subvarieties.*

Proof. Let N be a complex submanifold of a complex manifold M and let $p \in M$. If $p \in M \setminus N$, since N is closed in M, there is a neighborhood U of p which is disjoint from N and we may take the family $\mathcal{F}$ to consist of the constant function $f = 1$. Trivially, $N \cap U = \{q \in U : f(q) = 0\}$. Now, suppose $p \in N$. Then, by Theorem 51, there is an open neighborhood U of p in M and a holomorphic mapping $f : U \to \mathbb{C}^m$, such that $N \cap U = f^{-1}(0)$. That is, $N \cap U = \{q \in U : f_j(q) = 0, j = 1, \cdots, m\}$. Thus, N is a complex analytic subvariety of M. $\quad\square$

Analytic subvarieties are also called *analytic sets* (not to be confused with analytic sets in set theory). If we modify the definition of an analytic subvariety by requiring that it be locally the set of common zeros of polynomials rather than arbitrary holomorphic functions, then the analytic subvariety is called an *algebraic subvariety*. The subject of algebraic geometry is the study of algebraic varieties or equivalently the study of zeros of polynomials (or algebraic functions). In French, the subject of algebraic functions is géométrie algébrique and the subject of analytic functions is géométrie analytique. Thus, in French, géométrie analytique is several complex variables, the subject of these lecture notes, while in English analytic geometry is associated with calculus.

To understand analytic sets, it is essential to understand zero sets of holomorphic functions. The most fundamental result giving a local description of such zero sets is the Weierstrass preparation theorem 28. In closing, we suggest that the student review this basic result, equipped with the additional maturity he or she has attained subsequent to the first reading.

References

1. Bartle, R.G.: The Elements of Integration. Wiley, London (1966)
2. Bierstone, E., Milman, P.D., Parusinski, A.: A function which is arc-analytic but not continuous. Proc. Amer. Math. Soc. **113**(2), 419–423 (1991)
3. Conway, J.B.: Functions of One Complex Variable, 2nd edn. Springer, New York (1978)
4. Field, M.: Several Complex Variables and Complex Manifolds I. Cambridge University Press, Cambridge (1982)
5. Fleming, W.: Functions of Several Variables. 2nd edn. Undergraduate Texts in Mathematics. Springer, New York/Heidelberg (1977)
6. Gauthier, P.M., Ngô van Quê: Problème de surjectivité des applications holomorphes. Ann. Scuola Norm. Sup. Pisa (3) **27**, 555–559 (1973)
7. Kaplan, W.: Introduction to Analytic Functions. Addison-Wesley, Boston (1966)
8. Kaup, L., Kaup, B.: Holomorphic Functions of Several Variables. An Introduction to the Fundamental Theory. With the assistance of G. Barthel. Translated from the German by Michael Bridgland. de Gruyter Studies in Mathematics, vol. 3, Walter de Gruyter & Co., Berlin (1983)
9. Krantz, S.T.: Complex Analysis: The Geometric Viewpoint. Second edition. Carus Mathematical Monographs, vol. 23. Mathematical Association of America, Washington, DC (2004)
10. Merker, J., Porten, E.: A Morse-theoretical proof of the Hartogs extension theorem. J. Geom. Anal. **17**(3), 513–546 (2007)
11. Narasimhan, R.: Analysis on Real and Complex Manifolds. Reprint of the 1973 edition. North-Holland Mathematical Library, vol. 35, North-Holland Publishing Co., Amsterdam (1985)
12. Poincaré, H.: Les fonctions analytiques de deux variables et la représentation conforme. Rend. Circ. Mat. Palermo **23**, 185–220 (1907)
13. Range, R.M.: Holomorphic Functions and Integral Representations in Several Complex Variables. Graduate Texts in Mathematics, vol. 108, Springer, New York (1986), Second corrected printing (1998)
14. Range, R.M.: Some landmarks in the history of the tangential Cauchy Riemann equations. Rend. Mat. Appl. (7) **30**(3–4), 275–283 (2010)
15. Range, R.M.: What is ... a pseudoconvex domain? Notices Amer. Math. Soc. **59**(2), 301–303 (2012)

© Springer International Publishing Switzerland 2014
P.M. Gauthier, *Lectures on Several Complex Variables*,
DOI 10.1007/978-3-319-11511-5

16. Rudin, W.: Principles of Mathematical Analysis. 3rd edn. International Series in Pure and Applied Mathematics. McGraw-Hill Book Co., New York-Auckland-Düsseldorf (1976)
17. Rudin, W.: Function Theory in the Unit Ball of $\mathbb{C}^n$. Springer, New York (1980)
18. Wells, R.O., Jr.: Differential Analysis on Complex Manifolds. 3rd edn. With a new appendix by Oscar Garcia-Prada. Graduate Texts in Mathematics **65**, Springer, New York (2008)
19. Wermer, J.: On a domain equivalent to the bidisk. Math. Ann. **248**(3), 193–194 (1980)

Index

© Springer International Publishing Switzerland 2014   109
P.M. Gauthier, *Lectures on Several Complex Variables*,
DOI 10.1007/978-3-319-11511-5

Printed in the United States
By Bookmasters